COLLECTIVELY COMPACT OPERATOR APPROXIMATION THEORY

and

Applications to

Integral Equations

Prentice-Hall
Series in Automatic Computation
George Forsythe, editor

COLLECTIVELY COMPACT

OPERATOR

APPROXIMATION THEORY

and

Applications to

Integral Equations

PHILIP M. ANSELONE

Professor of Mathematics
Oregon State University
Corvallis, Oregon

With an Appendix by Joel Davis

PRENTICE-HALL, INC., Englewood Cliffs, New Jersey

Current printing (last digit):

10 9 8 7 6 5 4 3 2 1

13-140673-6

Library of Congress Catalog Card No.77-158108

Printed in the United States of America

PRENTICE-HALL INTERNATIONAL, INC., *London*
PRENTICE-HALL OF AUSTRALIA, PTY. LTD., *Sydney*
PRENTICE-HALL OF CANADA, LTD., *Toronto*
PRENTICE-HALL OF INDIA PRIVATE LIMITED, *New Delhi*
PRENTICE-HALL OF JAPAN, INC., *Tokyo*

To Joann and Cheryl

PREFACE

This book comprises an operator approximation theory and applications to the numerical solution of integral equations. It represents research published during the past few years by the author and several colleagues, principally Kendall E. Atkinson, Joel Davis, Robert H. Moore, and Theodore W. Palmer. Prerequisites for an understanding of the material include some acquaintance with functional analysis, integral equations, and numerical analysis. Many of the important concepts in these fields are explained at appropriate stages in the presentation. In addition, derivations of the basic properties of compact operators needed in the text are given in Appendix 1.

The setting for the operator approximation theory is an arbitrary Banach space. Most of the results pertain to bounded linear operators. The operator approximations converge pointwise rather than in operator norm, for which there is a classical and more elementary analysis. A *collective compactness* hypothesis effectively compensates for the discrepancy between pointwise and norm convergence. Principal results include convergence theorems and error bounds for operator inverses, plus approximation theorems for eigenvalues, eigenvectors, and spectral projections. Parts of the linear theory are extended to nonlinear operators with the aid of Fréchet differentiation and Newton's method.

The theoretical development is presented in Chapters 1 and 4–6. Chapters 2 and 3 give various applications to the approximate solution of integral equations by means of numerical integration. Special techniques are devised in order to handle more difficult types of problems, such as integral equations with discontinuous or singular kernels. The treatment of integral equations in the text is devoted primarily to the verification of the hypotheses of the abstract approximation theory. Numerical examples are presented and discussed in Appendix 2, which was written by Joel Davis. Readers interested primarily in integral equations are advised to look at the first few sections of Chapter 2 before beginning the development of the theory in Chapter 1.

There are, of course, many other methods for the approximate solution of integral and operator equations. Among the available references on this subject are the books by Bückner [29] and by Kantorovich and Krylov [48]. A forthcoming multivolume survey by Ben Noble entitled *Methods for Solving Integral Equations* will cover a considerable variety of numerical methods and give a number of applications to physics and engineering.

A limitation of many approximation methods for integral equations is that the theoretical approximations are expressed in terms of integrals which must be done numerically. Often the collectively compact operator approximation theory can be used to estimate the effect of the numerical integration errors. An example (cf. Chapter 3, §3.10) is furnished by the familiar technique of replacing the kernel of a Fredholm integral equation by a kernel of finite rank in order to derive an equation which is equivalent to a finite matrix problem. However, the matrix elements and the right members are integrals. Usually it is necessary to approximate these quantities by means of numerical integration. Since this is equivalent to the use of numerical integration directly on the derived integral equation of finite rank, the collectively compact theory is applicable. As another example, Newton's method for the approximate solution of nonlinear equations requires the solution of successive linear equations, which ordinarily must be attacked numerically. Once again (see Chapter 6) the collectively compact theory accounts for the effect of discretization.

Research on collectively compact sets of operators and their applications is currently under way in several directions. Some very interesting developments have occurred too late for inclusion in the book. A generalization of the theory to topological vector spaces has been initiated by J. D. DePree and J. A. Higgins [36]. The Ph.D. thesis of Higgins (New Mexico State University, 1971) will give simplified proofs of ergodic theorems which are based on collectively compact approximation theory, and will introduce the notion of a weakly collectively compact set of operators. Applications of the theory to boundary value problems, reformulated in terms of singular integral equations, have been made by K. E. Atkinson, D. L. Colton and R. P. Gilbert. A number of extensions of the ideas in this book are given by James [45], including the treatment of a very large class of integral operators with discontinuous kernels and generalizations to a topological vector space setting of the Korovkin theorem on the convergence of positive operators.

Most of my research included in this book was done at Oregon State University, under NSF and AEC contracts, and at the University of Wisconsin while a member of the Army Mathematics Research Center. Lecture series based on the material were presented at Oregon State University, Karlsruhe Technische Hochschule, Stanford University, and Michigan State University. A preliminary version of the book was prepared under an AEC grant at Stanford. My sincere appreciation goes to a number of colleagues, especially Kendall E. Atkinson, Joel Davis, Ronald B. Guenther, Jörg Hertling, Ralph L. James, John W. Lee, Robert H. Moore, Ben Noble, Theodore W. Palmer, and Richard S. Varga, for advice, encouragement, and for valuable criticisms of the manuscript. I am particularly indebted to Joel Davis for his contribution of the appendix on the numerical treatment of integral equations.

PHILIP M. ANSELONE
East Lansing, Michigan

CONTENTS

xi

chapter **3**

VARIANTS OF THE THEORY AND
FURTHER APPLICATIONS 33

chapter **4**

SPECTRAL APPROXIMATIONS 57

chapter **5**

CHARACTERIZATIONS OF COLLECTIVELY COMPACT
AND TOTALLY BOUNDED SETS OF OPERATORS 8I

Contents

COLLECTIVELY COMPACT OPERATOR APPROXIMATION THEORY

and

Applications to

Integral Equations

chapter I

AN APPROXIMATION THEORY FOR COMPACT LINEAR OPERATORS

I.I INTRODUCTION

The setting for the theory to be presented is an arbitrary, real or complex Banach space $\mathcal{X}$. Let $\mathcal{B} = \{x \in \mathcal{X} : \|x\| \leq 1\}$, the closed unit ball in $\mathcal{X}$. Denote by $[\mathcal{X}]$ the Banach space of bounded linear operators $T : \mathcal{X} \to \mathcal{X}$ with the usual operator norm, $\|T\| = \sup_{x \in \mathcal{B}} \|Tx\|$. In particular, I denotes the identity operator on $\mathcal{X}$. Convergence in norm of an operator sequence will be expressed by $\|T_n - T\| \to 0$ with the understanding that $n \to \infty$ through the positive integers. Pointwise convergence, $T_n x \to Tx$ for each $x \in \mathcal{X}$ as $n \to \infty$, will be indicated simply by $T_n \to T$.

Let $T \in [\mathcal{X}]$. The practical solution of an equation $Tx = y$ often involves operator approximations $T_n \in [\mathcal{X}]$, $n = 1, 2, \ldots$, such that the equations $T_n x_n = y$ can be solved by some means. For example, the latter equations may be equivalent to finite algebraic systems. Other features of a really satisfactory approximation scheme are:

T_n converges to T in an appropriate sense;

$T^{-1} \in [\mathcal{X}]$ and $T_n^{-1} \in [\mathcal{X}]$ for n sufficiently large;

1

T_n^{-1} converges to T^{-1};

$T_n^{-1} - T^{-1}$ is estimated conveniently in terms of $T_n - T$.

A standard theory of approximate solution, discussed briefly in §1.2, is based on the assumption that $\|T_n - T\| \to 0$. We shall impose the weaker hypothesis that $T_n \to T$. Certain compactness conditions will compensate for the inadequacy of pointwise convergence.

The material in this chapter is adapted from papers by Anselone and Moore [13] and Anselone [6].

I.2 OPERATOR NORM CONVERGENCE

The standard theory of approximate solution based on operator norm convergence is summarized for purposes of motivation, comparison, and later reference. The development begins with two elementary propositions which are stated without proofs.

PROPOSITION I.I. Let $A \in [\mathscr{X}]$ and $\|A\| < 1$. Then there exists $(I - A)^{-1} \in [\mathscr{X}]$,

$$(I - A)^{-1} = \sum_{m=0}^{\infty} A^m,$$

where the (Neumann) series converges in operator norm, and

$$\|(I - A)^{-1}\| \leq \frac{1}{1 - \|A\|}.$$

PROPOSITION I.2. Let $B, T \in [\mathscr{X}]$. If

$$BT = I - A, \qquad \|A\| < 1,$$

then T^{-1} exists (as an operator defined on $T\mathscr{X}$), and T^{-1} is bounded,

$$T^{-1} = (I - A)^{-1}B, \qquad T^{-1} - B = (I - A)^{-1}AB \quad \text{on} \quad T\mathscr{X},$$

$$\|T^{-1}\| \leq \frac{\|B\|}{1 - \|A\|}, \qquad \|T^{-1} - B\| \leq \frac{\|A\| \|B\|}{1 - \|A\|}.$$

Thus, if T has a sufficiently good approximate left inverse B, then T^{-1} exists and is near B. By a similar argument, if $TB = I - L$ with $\|L\| < 1$, then $T\mathcal{X} = \mathcal{X}$.

PROPOSITION 1.3. Let $S,T \in [\mathcal{X}]$. Assume there exists $S^{-1} \in [\mathcal{X}]$ and

$$\Delta = \|S^{-1}\| \, \|S - T\| < 1.$$

Then there exists $T^{-1} \in [\mathcal{X}]$ and

$$\|T^{-1}\| \leqq \frac{\|S^{-1}\|}{1 - \Delta}, \qquad \|T^{-1} - S^{-1}\| \leqq \frac{\Delta \, \|S^{-1}\|}{1 - \Delta}.$$

Proof. Let $B = S^{-1}$ in Proposition 1.2ff. Alternatively, a more traditional proof is based on the identities

$$T = S[I - S^{-1}(S - T)], \qquad T^{-1} - S^{-1} = T^{-1}(S - T)S^{-1},$$

and Proposition 1.1.

Proposition 1.3 implies that the operators $T \in [\mathcal{X}]$ with $T^{-1} \in [\mathcal{X}]$ form an open set and that the map $T \mapsto T^{-1}$ is continuous. In Proposition 1.3, replace S and T by T_n and T in both orders to obtain the following proposition.

PROPOSITION 1.4. Let $T, T_n \in [\mathcal{X}]$ and $\|T_n - T\| \to 0$. Then there exists $T^{-1} \in [\mathcal{X}]$ iff for some N and all $n \geq N$ there exist uniformly bounded $T_n^{-1} \in [\mathcal{X}]$, in which case $\|T_n^{-1} - T^{-1}\| \to 0$.

Bounds are available from Proposition 1.3.

1.3 COLLECTIVELY COMPACT SETS OF LINEAR OPERATORS

Since our analysis will depend heavily on compactness properties, a review of the basic concepts is in order. Some simplifications occur because $\mathcal{X}$ is complete.

Let $\mathscr{S} \subset \mathscr{X}$. Then $\mathscr{S}$ is *compact* iff each open cover of $\mathscr{S}$ has a finite subcover; $\mathscr{S}$ is *relatively compact* iff $\overline{\mathscr{S}}$ is compact. The set $\mathscr{S}$ is *sequentially compact* iff* each sequence in $\mathscr{S}$ has a convergent subsequence with the limit in $\mathscr{X}$. Finally, $\mathscr{S}$ is *totally bounded* iff for each $\varepsilon > 0$ $\mathscr{S}$ has a finite ε-net $\mathscr{S}_\varepsilon \subset \mathscr{X}$, i.e., corresponding to each $x \in \mathscr{S}$ there is an $x_\varepsilon \in \mathscr{S}_\varepsilon$ such that $\|x - x_\varepsilon\| < \varepsilon$. The following three properties are equivalent:

$\mathscr{S}$ is relatively compact,
$\mathscr{S}$ is sequentially compact,
$\mathscr{S}$ is totally bounded.

Total boundedness will be used in proofs of theorems much more often than either of the other equivalent concepts.

An operator $K \in [\mathscr{X}]$ is *compact* (or *completely continuous*) iff the set $K\mathscr{B}$ is relatively compact. We shall be concerned with equations of the form $(I - K)x = y$ with K compact. Special cases, considered in Chapter 2, are Fredholm integral equations of the second kind. According to classical theorems of Riesz on compact operators, $(I - K)^{-1}$ exists iff $(I - K)\mathscr{X} = \mathscr{X}$, in which case $(I - K)^{-1}$ is bounded and so belongs to $[\mathscr{X}]$ (cf. Appendix 1).

The following generalization of the notion of a compact operator will play a fundamental role in our approximation theory.

DEFINITION. A set $\mathscr{K} \subset [\mathscr{X}]$ is *collectively compact* provided that the set

$$\mathscr{K}\mathscr{B} = \{Kx\colon K \in \mathscr{K}, x \in \mathscr{B}\}$$

is relatively compact. A sequence of operators in $[\mathscr{X}]$ is collectively compact whenever the corresponding set is.

Clearly, every operator in a collectively compact set is compact. Any finite set of compact operators is collectively compact. Every subset of a collectively compact set is collectively compact. Further properties will be given when they are pertinent to the discussion.

---

* This agrees with Dunford and Schwartz [38] and also with the Russian school of functional analysts. Topologists usually require the limit to be in $\mathscr{S}$.

4

1.4 GENERAL HYPOTHESES

In this chapter we are concerned with operators $K, K_n \in [\mathscr{X}]$, $n = 1, 2, \ldots$, such that

(a) $K_n \to K$,
(b) $\{K_n\}$ is collectively compact,
(c) K is compact.

Actually, (a) and (b) imply (c), since

$$K\mathscr{B} \subset \overline{\{K_n x \colon n \geq 1, \quad x \in \mathscr{B}\}} = \overline{\{K_n\}\mathscr{B}}.$$

From (b), each operator K_n is compact.

It will be shown that $(I - K)^{-1}$ exists iff for some N and all $n \geq N$ the operators $(I - K_n)^{-1}$ exist and are bounded uniformly, in which case $(I - K_n)^{-1} \to (I - K)^{-1}$. Useful error estimates will be derived. These results will be applied in Chapters 2 and 3 to the approximate solution of integral equations by means of numerical integration. The theory resumes in Chapter 4 with a study of relations between spectral properties of operators $T, T_n \in [\mathscr{X}]$ which satisfy hypotheses weaker than (a) through (c), namely, $T_n \to T$ and $\{T_n - T\}$ is collectively compact.

1.5 POINTWISE CONVERGENCE OF OPERATOR INVERSES

The analysis begins with a few elementary observations. For $T \in [\mathscr{X}]$, define

$$m(T) = \inf_{\|x\|=1} \|Tx\| .$$

Then T^{-1} exists (as an operator on $T\mathscr{X}$) and is bounded iff $m(T) > 0$, in which case

$$\|T^{-1}\| = \frac{1}{m(T)} .$$

Given $S,T \in [\mathcal{X}]$, if S^{-1} and T^{-1} exist, then

$$S^{-1} - T^{-1} = S^{-1}(T - S)T^{-1} = T^{-1}(T - S)S^{-1} \quad \text{on} \quad S\mathcal{X} \cap T\mathcal{X}.$$

These remarks yield the following lemma.

LEMMA I.5. Let $T,T_n \in [\mathcal{X}]$, $n = 1, 2, \ldots$. If $T_n^{-1} \in [\mathcal{X}]$ and $\|T_n^{-1}\| \leq r < \infty$ for $n \geq N$, and if $T_n \to T$, then T^{-1} exists, $\|T^{-1}\| \leq r$, and $T_n^{-1} \to T^{-1}$ on $T\mathcal{X}$. If there is no N such that T_n^{-1} exists and is bounded uniformly for $n \geq N$, then there are infinite subsequences $\{T_{n_i}\}$ and $\{x_{n_i}\}$ such that

$$m(T_{n_i}) \to 0, \qquad T_{n_i} x_{n_i} \to 0, \qquad \|x_{n_i}\| = 1.$$

With this preparation we are ready to establish the following theorem on operator inverses.

THEOREM I.6. Let $K,K_n \in [\mathcal{X}]$, $n = 1, 2, \ldots$. Assume that $K_n \to K$, $\{K_n\}$ is collectively compact, and K is compact. Then $(I - K)^{-1}$ exists iff for some N and all $n \geq N$ the operators $(I - K_n)^{-1}$ exist and are bounded uniformly, in which case $(I - K_n)^{-1} \to (I - K)^{-1}$.

Proof. If there is such an N, the indicated conclusions follow from Lemma 1.5. If there is no such N, the lemma yields $\{K_{n_i}\}$ and $\{x_{n_i}\}$ such that

$$(I - K_{n_i})x_{n_i} \to 0, \qquad \|x_{n_i}\| = 1.$$

Since $\{K_n\}$ is collectively compact, $\{K_{n_i}\}\mathcal{B}$ is sequentially compact. Hence, there are subsequences and an element $x \in \mathcal{X}$ such that

$$K_{n_{i_j}} x_{n_{i_j}} \to x.$$

Then

$$x_{n_{i_j}} = K_{n_{i_j}} x_{n_{i_j}} + (I - K_{n_{i_j}})x_{n_{i_j}} \to x,$$

$$\|x\| = 1,$$

$$K_{n_{i_j}} x_{n_{i_j}} \to Kx,$$

$$(I - K)x = 0, \qquad x \neq 0.$$

So $(I - K)^{-1}$ does not exist and the proof is complete.

The foregoing reasoning may be familiar. In the special case with $K_n = K$ for all n, it becomes the usual contrapositive argument that $(I - K)^{-1}$ is bounded whenever $(I - K)^{-1}$ exists, which is part of the Riesz theory.

1.6 FROM POINTWISE TO NORM CONVERGENCE

Another proof of Theorem 1.6, which also yields error bounds, will be given a bit later. It will depend on the properties of pointwise convergence developed in this section.

PROPOSITION 1.7. Let $T, T_n \in [\mathcal{X}]$, $n = 1, 2, \ldots$, and $T_n \to T$. Then the sequence $\{T_n\}$ is bounded, and $T_n x \to Tx$ uniformly for x in any totally bounded set.

Proof. The first assertion is a special case of the Banach-Steinhaus theorem which, in turn, is a consequence of the principle of uniform boundedness. Let $\mathcal{S} \subset \mathcal{X}$ be totally bounded. Fix $\varepsilon > 0$. Then $\mathcal{S}$ has a finite ε-net $\mathcal{S}_\varepsilon$: for each $x \in \mathcal{S}$ there is an $x_\varepsilon \in \mathcal{S}_\varepsilon$ such that $\|x - x_\varepsilon\| < \varepsilon$. Hence, for $x \in \mathcal{S}$,

$$\|(T_n - T)x\| \leq \|(T_n - T)(x - x_\varepsilon)\| + \|(T_n - T)x_\varepsilon\|,$$

$$\|(T_n - T)x\| \leq (\|T_n\| + \|T\|)\varepsilon + \|(T_n - T)x_\varepsilon\|.$$

Since $T_n \to T$ and $\mathcal{S}_\varepsilon$ is finite, the desired uniform convergence follows.

The uniform convergence in Proposition 1.7 is a consequence of a more general fact. Recall that a linear operator T on $\mathcal{X}$ is bounded iff it is continuous. Analogously, a set of linear operators is bounded (uniformly) iff it is equicontinuous. Thus, a bounded sequence $\{T_n\}$ is equicontinuous. A proof similar to that given above shows that pointwise convergence of an equicontinuous sequence of functions from one metric space to another is uniform on any totally bounded set.

In contrast to pointwise convergence, $\|T_n - T\| \to 0$ iff $T_n x \to Tx$ uniformly for $x \in \mathcal{B}$ or, equivalently, for x in any bounded set. The bounded sets and the totally bounded sets coincide only when $\mathcal{X}$ is finite dimensional.

The following proposition will serve to bridge the gap between pointwise and norm convergence in the subsequent analysis.

PROPOSITION I.8. Let $T, T_n \in [\mathscr{X}]$, $n = 1, 2, \ldots,$ and $T_n \to T$. Then

$$\|(T_n - T)K\| \to 0$$

for each compact operator $K \in [\mathscr{X}]$. The convergence is uniform, with respect to K, for K in any collectively compact set $\mathscr{K} \subset [\mathscr{X}]$.

Proof. Let $\mathscr{K} \subset [\mathscr{X}]$ be collectively compact. Since $T_n \to T$ uniformly on the totally bounded set $\mathscr{K}\mathscr{B}$,

$$\|(T_n - T)Kx\| \to 0 \quad \text{uniformly for} \quad K \in \mathscr{K} \quad \text{and} \quad x \in \mathscr{B}.$$

The desired results follow.

COROLLARY I.9. Let $K, K_n \in [\mathscr{X}]$, $n = 1, 2, \ldots.$ Assume $K_n \to K$ and $\{K_n\}$ collectively compact. Then

$$\|(K_n - K)K\| \to 0, \qquad \|(K_n - K)K_n\| \to 0.$$

I.7 EXISTENCE AND APPROXIMATION OF OPERATOR INVERSES

A theorem on operator inverses designed to take advantage of Corollary 1.9 is established next; cf. Brakhage [25].

THEOREM I.10. Let $K, L \in [\mathscr{X}]$. Assume there exists $(I - K)^{-1} \in [\mathscr{X}]$ and

$$\Delta = \|(I - K)^{-1}(L - K)L\| < 1.$$

Then $(I - L)^{-1}$ exists and is bounded,

$$\|(I - L)^{-1}\| \leq \frac{1 + \|(I - K)^{-1}\| \, \|L\|}{1 - \Delta},$$

and

$$\|(I - L)^{-1}y - (I - K)^{-1}y\|$$

$$\leq \frac{\|(I - K)^{-1}\| \; \|Ly - Ky\| + \Delta \|(I - K)^{-1}y\|}{1 - \Delta}$$

for $y \in (I - L)\mathcal{X}$. (If L is compact then $(I - L)\mathcal{X} = \mathcal{X}$.)

Proof. The identity

$$(I - K)^{-1} = I + (I - K)^{-1}K$$

expresses $(I - K)^{-1}$ in terms of the so-called resolvent operator $(I - K)^{-1}K$. This suggests

$$B = I + (I - K)^{-1}L$$

as an approximate inverse for $I - L$. In effect, the resolvent operator is being approximated. A routine calculation yields

$$B(I - L) = I - A,$$

$$A = (I - K)^{-1}(L - K)L, \qquad \|A\| \leq \Delta < 1.$$

Proposition 1.2, with $T = I - L$, implies that $(I - L)^{-1}$ exists, is bounded, and

$$\|(I - L)^{-1}\| \leq \frac{\|B\|}{1 - \Delta}.$$

Another calculation yields

$$(I - L)^{-1} - (I - K)^{-1} = (I - A)^{-1}B - (I - K)^{-1}$$

$$= (I - A)^{-1}[B - (I - A)(I - K)^{-1}]$$

$$= (I - A)^{-1}[B - (I - K)^{-1} + A(I - K)^{-1}]$$

$$= (I - A)^{-1}[(I - K)^{-1}(L - K) + A(I - K)^{-1}]$$

on $(I - L)\mathcal{X}$. The theorem follows.

There is a generalization of Theorem 1.10 based on the condition

$$\Delta = \|(I - K)^{-1}\| \; \|(L - K)L^m\| < 1$$

for some integer $m \geq 1$. The identity

$$(I - K)^{-1} = I + K + \cdots + K^{m-1} + (I - K)^{-1}K^m$$

suggests

$$B = I + L + \cdots + L^{m-1} + (I - K)^{-1}L^m$$

as an approximate inverse of $I - L$. As before, Proposition 1.2 implies that $(I - L)^{-1}$ exists and

$$\|(I - L)^{-1}\| \leq \frac{\|B\|}{1 - \Delta}.$$

In the last inequality of Theorem 1.10, $Ly - Ky$ now should be replaced by $(L - K)(I + L + \cdots + L^{m-1})y$. The form of Δ and Corollary 1.9 indicate possible applications with L^m compact (cf. §1.8).

In a further generalization, due to Ostrowski [64], L^m is replaced by $p(L)$, where p is any polynomial such that $p(0) = 0$. This could be useful when $p(L)$ is compact.

I.8 CONVERGENCE THEOREMS AND ERROR BOUNDS

Theorem 1.10 will be applied to operators $K, K_n \in [\mathcal{X}]$ which satisfy the hypotheses in §1.4. To express the results in convenient and concise form, fix $y \in \mathcal{X}$ and, whenever the inverse operators exist, let

$$x = (I - K)^{-1}y, \qquad x_n = (I - K_n)^{-1}y.$$

THEOREM I.II. Let $K, K_n \in [\mathcal{X}]$, $n = 1, 2, \ldots$. Assume that $K_n \to K$, $\{K_n\}$ is collectively compact, and K is compact. Suppose $(I - K)^{-1}$ exists and define

$$\Delta_n = \|(I - K)^{-1}\| \, \|(K_n - K)K_n\|.$$

Then $\Delta_n \to 0$. Whenever $\Delta_n < 1$, $(I - K_n)^{-1}$ exists,

$$\|(I - K_n)^{-1}\| \leq \frac{1 + \|(I - K)^{-1}\| \, \|K_n\|}{1 - \Delta_n},$$

and

$$\|x_n - x\| \leq \frac{\|(I - K)^{-1}\| \, \|K_n y - Ky\| + \Delta_n \|x\|}{1 - \Delta_n} \to 0.$$

The estimates for $\|(I - K_n)^{-1}\|$ are bounded uniformly in n.

Proof. Let $L = K_n$ in Theorem 1.10 and refer to Corollary 1.9.

Theorem 1.11 yields a direct proof of the statement which was proved indirectly in Theorem 1.6, plus a criterion for the number N and pertinent inequalitites. Other inequalities follow:

$$\|x_n\| \leq \left(\frac{1 + \|(I - K)^{-1}\| \, \|K_n\|}{1 - \Delta_n} \right) \|y\| ,$$

$$\|x_n - x\| \leq \frac{\|(I - K)^{-1}\| \, (\|K_n y - Ky\| + \Delta_n \|y\|)}{1 - \Delta_n} \to 0.$$

This bound for $\|x_n - x\|$ involves y but neither x nor x_n, which is advantageous when a set of elements y is being considered. Theorem 1.11 and the identity

$$x_n - x = (I - K_n)^{-1}[y - (I - K_n)x]$$

yield still another estimate:

$$\|x_n - x\| \leq \frac{(1 + \|(I - K)^{-1}\| \, \|K_n\|) \, \|y - (I - K_n)x\|}{1 - \Delta_n} \to 0.$$

The next result is analogous to Theorem 1.11, but with the roles of K and K_n interchanged.

THEOREM 1.12. Let $K, K_n \in [\mathscr{X}]$, $n = 1, 2, \ldots$. Assume that $K_n \to K$, $\{K_n\}$ is collectively compact, and K is compact. Whenever $(I - K_n)^{-1}$ exists, define

$$\Delta^n = \|(I - K_n)^{-1}\| \, \|(K_n - K)K\| .$$

For a particular n assume that $(I - K_n)^{-1}$ exists and $\Delta^n < 1$. Then $(I - K)^{-1}$ exists,

$$\|(I - K)^{-1}\| \leq \frac{1 + \|(I - K_n)^{-1}\| \, \|K\|}{1 - \Delta^n} ,$$

and

$$\|x_n - x\| \leq \frac{\|(I - K_n)^{-1}\| \, \|K_n y - Ky\| + \Delta^n \|x_n\|}{1 - \Delta^n} .$$

Moreover, $(I - K_n)^{-1}$ exists for all n sufficiently large, $\Delta^n \to 0$, the estimates for $\|(I - K)^{-1}\|$ are bounded uniformly with respect to n, and the estimates for $\|x_n - x\|$ approach zero as $n \to \infty$.

11

Proof. In Theorem 1.10 replace K and L by K_n and K, respectively. Then use Theorem 1.11 and Corollary 1.9.

Note that Theorem 1.12 gives the existence of $(I - K)^{-1}$ and related inequalities under conditions on K_n for a single n. This is of considerable value in practical applications. Theorem 1.12 also yields

$$\|x\| \leq \left(\frac{1 + \|(I - K_n)^{-1}\| \, \|K\|}{1 - \Delta^n} \right) \|y\| ,$$

$$\|x_n - x\| \leq \frac{\|(I - K_n)^{-1}\| \, (\|K_n y - Ky\| + \Delta^n \|y\|)}{1 - \Delta^n} \to 0,$$

$$\|x_n - x\| \leq \frac{(I + \|(I - K_n)^{-1}\| \, \|K\|) \, \|y - (I - K)x_n\|}{1 - \Delta^n} \to 0.$$

Since the bounds in Theorem 1.12 and following text involve $(I - K_n)^{-1}$ and x_n but neither $(I - K)^{-1}$ nor x, they are particularly suitable for practical error estimation.

1.9   CONCLUDING REMARKS

The theory just developed has an immediate extension expressed in terms of $\lambda I - K$ and $\lambda_n I - K_n$, where $\lambda_n \to \lambda \neq 0$. For example, $\lambda_n = \lambda \neq 0$ for all n. One merely replaces K and K_n by K/λ and K_n/λ_n in the previous analysis. A much more elaborate spectral approximation theory is given in Chapter 4. Prior to that, we give a number of applications of the preceding material to integral equations.

In a forthcoming paper entitled "A Characterization of Collectively Compact Sets of Linear Operators," J. A. Higgins establishes a number of interesting results. We cite two of them. A set $\mathcal{K} \subset [\mathcal{X}]$ is collectively compact iff each sequence $\{K_n\} \subset \mathcal{K}$ is collectively compact. A sequence $\{K_n\} \subset \mathcal{K}$ is collectively compact iff $\{K_n x_n\}$ is relatively compact for each sequence $\{x_n\} \subset \mathcal{B}$.

chapter 2

APPROXIMATE SOLUTIONS OF INTEGRAL EQUATIONS

2.1 INTRODUCTION AND SUMMARY

Consider a Fredholm integral equation

$$x(s) - \int_0^1 k(s,t)x(t)\, dt = y(s), \qquad 0 \le s \le 1.$$

To begin with, the discussion is informal. Precise conditions are specified later.

Numerical integration with a general quadrature formula yields equations

$$x_n(s) - \sum_{j=1}^{n} w_{nj}k(s,t_{nj})x_n(t_{nj}) = y(s), \qquad 0 \le s \le 1, \qquad n = 1, 2, \ldots,$$

for approximate solutions of the integral equation. For each n the preceding equation provides an interpolation formula for $x_n(s)$ in terms of $x_n(t_{nj})$, $j = 1, \ldots, n$. Therefore, the equation for $x_n(s)$ reduces to the finite algebraic system

$$x_n(t_{ni}) - \sum_{j=1}^{n} w_{nj}k(t_{ni},t_{nj})x_n(t_{nj}) = y(t_{ni}), \qquad i = 1, \ldots, n.$$

13

Although other interpolation formulas are often used, the given equation for $x_n(s)$ has the important theoretical advantage that the equations for $x(s)$ and $x_n(s)$ have the same general form:

$$(I - K)x = y, \qquad (I - K_n)x_n = y.$$

The integral operator K in the former is approximated by a numerical-integral operator K_n in the latter. This will enable us to apply the approximation theory developed in Chapter 1.

The approximate solution of integral equations by means of numerical integration goes back to at least Fredholm [40, 41]. Convergence questions were studied by Hilbert [42]. Apparently, Nyström [63] was the first to employ the above equation for $x_n(s)$ in particular cases. Convergence theorems and error bounds for approximate solutions have been obtained under various assumptions by a number of investigators, including Brakhage [25, 26], Bückner [28–30], Fox and Goodwin [39], Kantorovich and Krylov [48], and Mysovskih [54–56].

The contributions of Brakhage, Kantorovich and Krylov, and Mysovskih are direct historical antecedents of the approximation theory presented here. They obtained some of the same convergence results and error bounds in special cases. A perusal of those bounds led eventually to the concept of a collectively compact set of operators and to the related abstract approximation theory. That theory will be applied to a wide variety of integral equations in Chapters 2 and 3.

Initially the analysis will be carried out in the space $\mathscr{C}$ of real or complex continuous functions $x(t)$, $0 \leq t \leq 1$, with the uniform norm, $\|x\| = \max |x(t)|$. Convergence in norm is uniform convergence, which is advantageous for numerical calculations. The analysis simplifies somewhat when the functions are real.

As for the quadrature formula, it will be assumed that

$$\sum_{j=1}^{n} w_{nj}x(t_{nj}) \rightarrow \int_0^1 x(t)\, dt, \qquad x \in \mathscr{C}.$$

The rectangular, trapezoidal, Simpson, Weddle and Gauss quadrature rules satisfy this condition, but the Newton-Cotes formulas do not (cf. Hildebrand [43], Milne [53]).

When the kernel $k(s,t)$ is continuous for $0 \leq s,t \leq 1$, the integral operator K and the corresponding numerical-integral operators K_n belong to $[\mathscr{C}]$. Then (cf. §2.3) K is compact, $\{K_n\}$ is collectively compact, and $K_n \rightarrow K$,

14

so the general theory is applicable. Computability of the error bounds will be discussed later. Numerical examples are given in Appendix 2 written by Joel Davis.

A compact integral operator $K \in [\mathscr{C}]$ may have a discontinuous kernel. But then the corresponding numerical-integral operators ordinarily fail to map $\mathscr{C}$ in $\mathscr{C}$. To deal with that situation, K and K_n will be defined on the larger Banach space $\mathscr{R}$ of properly Riemann integrable functions $x(t)$, $0 \le t \le 1$, with the supremum norm. For a large class of discontinuous kernels, it will be shown that $K_n \to K$ and $\{K_n\}$ is collectively compact, so the general theory applies. Special cases include Volterra kernels $k(s,t)$ which are continuous for $0 \le t \le s \le 1$ and zero for $0 \le s < t \le 1$.

The effectiveness of the approximation theory applied to integral equations with discontinuous kernels is limited because convergence of numerical integrals of discontinuous functions is usually slow. Fortunately, it is often possible to modify the approximation method so as to mitigate the adverse effects of discontinuities or even mild singularities in the kernels. Such techniques are described in Chapter 3 along with other variations of the ideas in the first two chapters.

The material in this chapter pertaining to integral operators with continuous kernels is based primarily on references [13] by Anselone and Moore and [6] by Anselone. The treatment of integral operators on $\mathscr{R}$ with discontinuous kernels is adapted from [7, 8] by Anselone.

2.2   THE QUADRATURE FORMULA ON $\mathscr{C}$

Define linear functionals φ and φ_n, $n = 1, 2, \ldots$, on $\mathscr{C}$ by

$$\varphi x = \int_0^1 x(t)\, dt, \qquad \varphi_n x = \sum_{j=1}^n w_{nj} x(t_{nj}),$$

where $0 \le t_{nj} \le 1$ and the "weights" w_{nj} are real or complex. Then φ and φ_n are bounded and

$$\|\varphi\| = 1, \qquad \|\varphi_n\| = \sum_{j=1}^n |w_{nj}|.$$

Assume that

$$\varphi_n \to \varphi,$$

i.e., $\varphi_n x \to \varphi x$ for each $x \in \mathscr{C}$ as $n \to \infty$. By the Banach-Steinhaus theorem, the sequence $\{\varphi_n\}$ is bounded. Thus,

$$\|\varphi_n\| = \sum_{j=1}^{n} |w_{nj}| \leq b < \infty$$

for some b and all n.

Define $u \in \mathscr{C}$ by $u(t) \equiv 1$. Then $\varphi_n u \to \varphi u$ implies

$$\sum_{j=1}^{n} w_{nj} \to 1.$$

For most of the common quadrature formulas, every $w_{nj} \geq 0$. In that case

$$\|\varphi_n\| = \sum_{j=1}^{n} w_{nj} \to 1,$$

so $\{\varphi_n\}$ is bounded without recourse to the Banach-Steinhaus theorem. Often, every $w_{nj} \geq 0$ and $\varphi_n u = \varphi u = 1$ for all n. Then $\|\varphi_n\| = 1$ for all n, and $b = 1$ suffices as a bound for $\{\varphi_n\}$.

The pointwise convergence $\varphi_n \to \varphi$ is uniform on the totally bounded sets in $\mathscr{C}$. These are the bounded equicontinuous sets according to the Arzelà-Ascoli theorem. If $\varphi_n u = \varphi u$ for all n, then $\varphi_n \to \varphi$ uniformly on each equicontinuous set $\mathscr{F} \subset \mathscr{C}$, whether bounded or not, because the set

$$\{x_0 = x - x(0)u : x \in \mathscr{F}\}$$

is bounded and equicontinuous, and

$$\varphi_n x - \varphi x = \varphi_n x_0 - \varphi x_0, \qquad x \in \mathscr{F}, \qquad n \geq 1.$$

To illustrate the foregoing uniform convergence, let $\mathscr{F}$ be the set of functions $x \in \mathscr{C}$ which satisfy a uniform Lipschitz condition

$$|x(s) - x(t)| \leq M |s - t|.$$

Clearly, $\mathscr{F}$ is equicontinuous. For the rectangular quadrature formula,

$$\varphi_n x = \frac{1}{n} \sum_{j=1}^{n} x\left(\frac{j}{n}\right),$$

it is easy to prove that

$$|\varphi_n x - \varphi x| \leq \frac{M}{2n}, \qquad x \in \mathscr{F}, \qquad n \geq 1.$$

For the repeated midpoint rule,

$$\varphi_n x = \frac{1}{n} \sum_{j=1}^{n} x\left(\frac{j - \frac{1}{2}}{n}\right),$$

we have

$$|\varphi_n x - \varphi x| \leq \frac{M}{4n}, \qquad x \in \mathscr{F}, \qquad n \geq 1.$$

For the trapezoidal rule with $n + 1$ points,

$$\varphi_{n+1} x = \frac{1}{n}\left[\tfrac{1}{2}x(0) + \tfrac{1}{2}x(n) + \sum_{j=1}^{n-1} x\left(\frac{j}{n}\right)\right],$$

the error formula is

$$|\varphi_{n+1} x - \varphi x| \leq \frac{M}{4n}, \qquad x \in \mathscr{F}, \qquad n \geq 1.$$

Hence, in all three cases, $\varphi_n \to \varphi$ uniformly on $\mathscr{F}$.

If $\mathscr{F}$ is the set of differentiable functions $x \in \mathscr{C}$ such that

$$|x'(s) - x'(t)| \leq M\,|s - t|,$$

then for the repeated midpoint rule we have

$$|\varphi_n x - \varphi x| \leq \frac{M}{24n^2}, \qquad x \in \mathscr{F}, \qquad n \geq 1,$$

and for the trapezoidal rule we have

$$|\varphi_{n+1} x - \varphi x| \leq \frac{M}{12n^2}, \qquad x \in \mathscr{F}, \qquad n \geq 1.$$

Again, $\varphi_n \to \varphi$ uniformly on $\mathscr{F}$, although now $\mathscr{F}$ is neither bounded nor equicontinuous. In a similar fashion, error bounds expressed in terms of bounds on derivatives of some order imply uniform convergence on certain sets which are neither bounded nor equicontinuous. Such bounds are associated with quadrature formulas which are exact on polynomials of particular degrees.

In spite of the preceding remarks the convergence of φ_n to φ is not uniform on all the bounded sets. Equivalently,

$$\|\varphi_n - \varphi\| \not\to 0.$$

To verify this, construct piecewise linear functions $x_n \in \mathscr{C}$ such that $0 \leq x_n \leq 1$, $x_n(t_{nj}) = 0$ for $1 \leq j \leq n$, and $\varphi x_n \to 1$. Then $\varphi_n x_n = 0$ for all n, $\varphi_n x_n - \varphi x_n \to -1$ and, hence, $\|\varphi_n - \varphi\| \not\to 0$.

17

2.3 APPROXIMATIONS OF INTEGRAL OPERATORS WITH CONTINUOUS KERNELS

Continue the notation and assumptions of §2.2. Thus, in particular, $\varphi_n \to \varphi$. Define operators $K, K_n \in [\mathscr{C}]$ for $n = 1, 2, \ldots$ by

$$(Kx)(s) = \int_0^1 k(s,t)x(t)\, dt,$$

$$(K_n x)(s) = \sum_{j=1}^n w_{nj} k(s,t_{nj}) x(t_{nj}),$$

where $k(s,t)$ is continuous for $0 \leqq s,t \leqq 1$. Let

$$\|k\| = \max |k(s,t)|.$$

Standard arguments yield

$$\|K\| = \max_{0 \leqq s \leqq 1} \int_0^1 |k(s,t)|\, dt \leqq \|k\|,$$

$$\|K_n\| = \max_{0 \leqq s \leqq 1} \sum_{j=1}^n |w_{nj} k(s,t_{nj})| \leqq b\,\|k\|.$$

Thus the sequence $\{K_n\}$ is bounded.

For convenience below, define $k_s, k^t \in \mathscr{C}$ by

$$k_s(t) = k^t(s) = k(s,t), \qquad 0 \leqq s,t \leqq 1.$$

Since k is uniformly continuous, the sets

$$\{k_s : 0 \leqq s \leqq 1\}, \qquad \{k^t : 0 \leqq t \leqq 1\}$$

are bounded and equicontinuous. Observe that

$$(Kx)(s) = \varphi(k_s x), \qquad (K_n x)(s) = \varphi_n(k_s x).$$

Since $\varphi_n \to \varphi$ uniformly on the bounded equicontinuous set $\{k_s x : 0 \leqq s \leqq 1\}$ for each $x \in \mathscr{C}$, we have the following proposition.

PROPOSITION 2.1. $K_n \to K$.

However, $\|K_n - K\| \to 0$ iff $K = O$. In fact, $\|K_n - K\| \to 2\,\|K\|$. This can be shown by a construction similar to that for $\|\varphi_n - \varphi\| \nrightarrow 0$.

18

PROPOSITION 2.2. K is compact and $\{K_n\}$ is collectively compact.

Proof. The argument for K is well known. It is based on the inequalities

$$\|Kx\| \leq \|K\| \, \|x\|,$$

$$|(Kx)(s) - (Kx)(s')| \leq \|k_s - k_{s'}\| \, \|x\|$$

and the Arzelà-Ascoli theorem. Similarly, the inequalities

$$\|K_n x\| \leq \|K_n\| \, \|x\| \leq b \, \|k\| \, \|x\|,$$

$$|(K_n x)(s) - (K_n x)(s')| \leq b \, \|k_s - k_{s'}\| \, \|x\|$$

imply that $\{K_n\}$ is collectively compact. (Hence, each K_n is compact; this also follows from the fact that $\dim K_n \mathscr{C} \leq n < \infty$.)

In view of Propositions 2.1 and 2.2, the approximation theory of Chapter 1 applies to the equations

$$(I - K)x = y, \qquad (I - K_n)x_n = y.$$

We call particular attention to Theorems 1.6, 1.11 and 1.12.

2.4   DETERMINATION OF THE APPROXIMATE SOLUTIONS

As we observed in §2.1, for each n the equation $(I - K_n)x_n = y$ reduces to a finite algebraic system. To make this more precise we utilize the vector space l_n^∞. The vectors $v = (v_1, \ldots, v_n)$ have real or complex components and the norm is $\|v\| = \max |v_j|$. Let I_n be the identity operator on l_n^∞. Define a bounded linear operator $P_n \colon \mathscr{C} \to l_n^\infty$ by

$$P_n x = v, \qquad v_j = x(t_{nj}).$$

Then $\|P_n\| = 1$ and $P_n \mathscr{C} = l_n^\infty$.

Define bounded linear operators $\hat{K}_n \colon l_n^\infty \to \mathscr{C}$ and $\tilde{K}_n \colon l_n^\infty \to l_n^\infty$ by

$$(\hat{K}_n v)(s) = \sum_{j=1}^{n} w_{nj} k(s, t_{nj}) v_j,$$

$$(\tilde{K}_n v)_i = \sum_{j=1}^{n} w_{nj} k(t_{ni}, t_{nj}) v_j.$$

Thus, $\tilde{K}_n$ has the matrix representation $[w_{nj}k(t_{ni},t_{nj})]$. The norm of K_n is the *maximum absolute row-sum*,

$$\|\tilde{K}_n\| = \max_i \sum_{j=1}^{n} |w_{nj}k(t_{ni},t_{nj})|\,.$$

It is not difficult to verify that

$$\|\tilde{K}_n\| \leq \|\hat{K}_n\| = \|K_n\|\,.$$

The operators K_n, $\hat{K}_n$, $\tilde{K}_n$ and P_n are related by

$$K_n = \hat{K}_n P_n, \qquad \tilde{K}_n = P_n \hat{K}_n, \qquad P_n K_n = \tilde{K}_n P_n = P_n \hat{K}_n P_n.$$

It follows that

$$(I - K_n)x_n = y, \qquad v_n = P_n x_n \qquad \Rightarrow \qquad (I_n - \tilde{K}_n)v_n = P_n y.$$

Conversely,

$$(I_n - \tilde{K}_n)v_n = P_n y, \qquad x_n = y + \hat{K}_n v_n \qquad \Rightarrow$$

$$P_n x_n = P_n y + P_n \hat{K}_n v_n = P_n y + \tilde{K}_n v_n = v_n,$$

$$K_n x_n = \hat{K}_n P_n x_n = \hat{K}_n v_n,$$

$$(I - K_n)x_n = y.$$

These considerations yield the following proposition.

PROPOSITION 2.3. There exists $(I - K_n)^{-1} \in [\mathscr{C}]$ iff there exists $(I_n - \tilde{K}_n)^{-1} \in [l_n^\infty]$, in which case

$$(I - K_n)^{-1} = I + \hat{K}_n(I_n - \tilde{K}_n)^{-1}P_n,$$

$$\|(I - K_n)^{-1}\| \leq 1 + \|K_n\|\,\|(I_n - \tilde{K}_n)^{-1}\|.$$

Thus, $\|(I - K_n)^{-1}\|$ can be estimated numerically in terms of $\|K_n\|$ and the matrix norm $\|(I_n - \tilde{K}_n)^{-1}\|$ which is a maximum absolute row-sum.

As we have seen, $(I - K_n)x_n = y$ is essentially equivalent to the $n \times n$ matrix problem $(I_n - \tilde{K}_n)P_n x_n = P_n y$. In §3.7 of Chapter 3, techniques are given which circumvent the necessity of dealing with matrices of prohibitively large size.

20

2.5 ERROR BOUNDS

Suppose that $(I - K)x = y$ and $(I - K_n)x_n = y$. The estimates for $\|x_n - x\|$ in Chapter 1 depend critically on $\|K_n y - Ky\|$, $\|(K_n - K)K\|$, and $\|(K_n - K)K_n\|$. From the analysis in §2.3,

$$\|K_n y - Ky\| = \max_{0 \leq s \leq 1} |(\varphi_n - \varphi)(k_s y)| .$$

Quadrature error formulas yield numerical estimates for $\|K_n y - Ky\|$ when k and y have sufficient smoothness properties.

Define

$$k_2(s,t) = \int_0^1 k(s,\tau)k(\tau,t)\, d\tau = \varphi(k_s k^t),$$

$$k_{n2}(s,t) = \sum_{j=1}^n w_{nj} k(s,t_{nj})k(t_{nj},t) = \varphi_n(k_s k^t).$$

Then

$$(K^2 x)(s) = \int_0^1 k_2(s,t)x(t)\, dt,$$

$$(K_n K x)(s) = \int_0^1 k_{n2}(s,t)x(t)\, dt,$$

$$(K K_n x)(s) = \sum_{j=1}^n w_{nj} k_2(s,t_{nj})x(t_{nj}),$$

$$(K_n^2 x)(s) = \sum_{j=1}^n w_{nj} k_{n2}(s,t_{nj})x(t_{nj}),$$

and

$$\|(K_n - K)K\| = \max_{0 \leq s \leq 1} \int_0^1 |k_{n2}(s,t) - k_2(s,t)|\, dt,$$

$$\|(K_n - K)K_n\| = \max_{0 \leq s \leq 1} \sum_{j=1}^n |w_{nj}|\, |k_{n2}(s,t_{nj}) - k_2(s,t_{nj})| .$$

It follows that

$$\|(K_n - K)K\| \leq \delta_n, \qquad \|(K_n - K)K_n\| \leq b\, \delta_n,$$

where

$$\delta_n = \max_{0 \leq s,t \leq 1} |(\varphi_n - \varphi)(k_s k^t)| ,$$

and $\delta_n \to 0$ since $\varphi_n \to \varphi$ uniformly on the bounded equicontinuous set $\{k_s k^t: 0 \leq s,t \leq 1\}$. Hence, in agreement with the abstract theory,

$$\|(K_n - K)K\| \to 0, \qquad \|(K_n - K)K_n\| \to 0.$$

Quadrature error formulas yield numerical estimates for these quantities if k is smooth enough.

The results in §2.3–§2.5 and the inequalities in Theorems 1.11 and 1.12 yield computable estimates for $\|x_n - x\|$ which tend to zero as $n \to \infty$ when k and y are at least Lipschitz continuous. The rate of convergence depends on the smoothness of the functions. For numerical examples see Appendix 2 by Joel Davis.

2.6 THE BANACH SPACE $\mathscr{R}$

In order to deal with compact integral operators having discontinuous kernels, we introduce the linear space $\mathscr{R}$ of (properly) Riemann integrable, real or complex functions $x(t)$, $0 \leq t \leq 1$, equipped with the norm $\|x\| = \sup |x(t)|$. Recall that a function x defined on [0, 1] belongs to $\mathscr{R}$ iff x is continuous almost everywhere and is bounded. It follows by an easy argument that $\mathscr{R}$ is complete. Thus, $\mathscr{R}$ is a Banach space. Clearly, $\mathscr{C}$ is a closed subspace of $\mathscr{R}$.

Since pointwise convergence of bounded linear operators is uniform on totally bounded sets, we shall need characterizations of totally bounded sets in $\mathscr{R}$. For example, any bounded equicontinuous set is totally bounded in $\mathscr{C}$, hence in $\mathscr{R}$. A related class of totally bounded sets will be defined in terms of the following concept.

DEFINITION. A sequence $\{\mathscr{F}_n\}$ of sets in $\mathscr{R}$ is *asymptotically equicontinuous* if

$$\sup_{x \in \mathscr{F}_n} |x(s) - x(t)| \to 0 \quad \text{as} \quad \begin{matrix} s \to t, \\ n \to \infty, \end{matrix} \quad \text{uniformly for} \quad t \in [0,1];$$

equivalently, for each $\varepsilon > 0$ there exist $\delta = \delta(\varepsilon) > 0$ and $N = N(\varepsilon) \geq 1$ such that

$$|x(s) - x(t)| < \varepsilon \quad \text{for} \quad |s - t| < \delta(\varepsilon), \qquad x \in \mathscr{F}_n, \qquad n \geq N.$$

PROPOSITION 2.4. Let $\{\mathscr{F}_n\}$ be a uniformly bounded asymptotically equicontinuous sequence of totally bounded sets in $\mathscr{R}$. Then $\bigcup \mathscr{F}_n$ is totally bounded.

Proof. Fix $\varepsilon > 0$. Choose $\delta = \delta(\varepsilon)$ and $N = N(\varepsilon)$ as in the preceding definition. A construction very similar to that used in the usual proof of the Arzelà-Ascoli theorem yields a finite ε-net for $\bigcup_{n \geq N} \mathscr{F}_n$ consisting of step functions. Since each $\mathscr{F}_n$ is totally bounded, $\bigcup_{n < N} \mathscr{F}_n$ has a finite ε-net. It follows that $\bigcup_{n \geq 1} \mathscr{F}_n$ is totally bounded.

Roughly speaking, the sets $\mathscr{F}_n$ in an asymptotically equicontinuous sequence become more and more nearly equicontinuous as n increases. This remark has to be interpreted rather loosely because the functions involved may have discontinuities—which decrease in magnitude as $n \to \infty$.

For example, let $\mathscr{F}$ be a bounded equicontinuous set in $\mathscr{R}$ and, corresponding to each $x \in \mathscr{F}$, define a sequence $\{x_n\}$ in $\mathscr{R}$ such that $\|x_n - x\| \to 0$ uniformly for $x \in \mathscr{F}$ as $n \to \infty$. By the triangle inequality,

$$|x_n(s) - x_n(t)| \leq 2\,\|x_n - x\| + |x(s) - x(t)|.$$

Let $\mathscr{F}_n = \{x_n : x \in \mathscr{F}\}$. Then $\{\mathscr{F}_n\}$ is uniformly bounded and asymptotically equicontinuous. If each $\mathscr{F}_n$ is totally bounded, then $\bigcup \mathscr{F}_n$ is totally bounded by Proposition 2.4.

As a more concrete illustration, let $\mathscr{F}$ be a bounded equicontinuous set of real functions in $\mathscr{R}$. For each $x \in \mathscr{F}$ construct step function approximations x_n, $n = 1, 2, \ldots$, such that both the points of discontinuity and the values of x_n are integral multiples of $1/n$ and

$$\left| x_n\!\left(\frac{m}{n}\right) - x\!\left(\frac{m}{n}\right) \right| < \frac{1}{n}, \qquad m = 0, 1, \ldots, n.$$

Then $\|x_n - x\| \to 0$ uniformly for $x \in \mathscr{F}$ as $n \to \infty$. Let $\mathscr{F}_n = \{x_n : x \in \mathscr{F}\}$. As before, $\{\mathscr{F}_n\}$ is uniformly bounded and asymptotically equicontinuous. Each $\mathscr{F}_n$ is a finite set, hence totally bounded. So $\bigcup \mathscr{F}_n$ is totally bounded.

2.7   THE QUADRATURE FORMULA ON $\mathscr{R}$

Let φ and φ_n be bounded linear functionals on $\mathscr{R}$ of the forms

$$\varphi x = \int_0^1 x(t)\, dt, \qquad \varphi_n x = \sum_{j=1}^n w_{nj} x(t_{nj}), \qquad n = 1, 2, \ldots.$$

Assume

$$\varphi_n x \to \varphi x, \qquad x \in \mathscr{C},$$

$$w_{nj} \geqq 0, \qquad 1 \leqq j \leqq n.$$

These conditions usually are satisfied in practice. As in §2.2,

$$\|\varphi\| = 1, \qquad \|\varphi_n\| = \sum_{j=1}^{n} w_{nj} \leqq b$$

for some $b < \infty$ and all n.

Note that φx and $\varphi_n x$ are real when x is real and, for any $x \in \mathscr{R}$,

$$\varphi x = \varphi(\operatorname{Re} x) + i\varphi(\operatorname{Im} x),$$

$$\varphi_n x = \varphi_n(\operatorname{Re} x) + i\varphi_n(\operatorname{Im} x).$$

Consequently, many assertions reduce to the corresponding real cases.

For arbitrary complex numbers, let $z_1 \geqq z_2$ and $z_2 \leqq z_1$ whenever $\operatorname{Re} z_1 \geqq \operatorname{Re} z_2$ and $\operatorname{Im} z_1 \geqq \operatorname{Im} z_2$. For x and y in $\mathscr{R}$, let

$$x \geqq 0 \quad \text{and} \quad 0 \leqq x \quad \text{whenever} \quad x(t) \geqq 0 \qquad \text{for} \quad 0 \leqq t \leqq 1,$$

$$x \geqq y \quad \text{and} \quad y \leqq x \quad \text{whenever} \quad x(t) \geqq y(t) \quad \text{for} \quad 0 \leqq t \leqq 1.$$

Then

$$x \geqq 0 \quad \Rightarrow \quad \varphi x \geqq 0 \quad \text{and} \quad \varphi_n x \geqq 0,$$

$$x \geqq y \quad \Rightarrow \quad \varphi x \geqq \varphi y \quad \text{and} \quad \varphi_n x \geqq \varphi_n y.$$

These positivity and monotonicity properties will play key roles in the analysis.

The upper and lower approximating sums for the Riemann integral of a real function can be interpreted as integrals of step functions. Slight modifications make them integrals of continuous functions. Thus, the following relationship between $\mathscr{C}$ and $\mathscr{R}$ is easily established.

LEMMA 2.5. A complex function $x(t)$, $0 \leqq t \leqq 1$, is in $\mathscr{R}$ iff for $m = 1, 2, \ldots$ there exist $x_m, x^m \in \mathscr{C}$ such that

$$x_m \leqq x \leqq x^m,$$

$$\varphi(x^m - x_m) \to 0,$$

in which case

$$\varphi x = \lim_{m \to \infty} \varphi x_m = \lim_{m \to \infty} \varphi x^m.$$

If x is real (resp., nonnegative), then x_m and x^m may be assumed to be real (resp., nonnegative).

A similar result, with $\mathscr{C}$ replaced by the polynomials on $[0, 1]$, is due to Weyl [73].

For each $x \in \mathscr{R}$ let

$$\|x\|_1 = \varphi(|x|) = \int_0^1 |x(t)| \, dt.$$

This defines the $\mathscr{L}^1$ seminorm on $\mathscr{R}$. It fails to be a norm because $\|x\|_1 = 0 \not\Rightarrow x = 0$. Under the conditions in Lemma 2.5, $\|x_m - x\|_1 \to 0$ and $\|x^m - x\|_1 \to 0$ as $m \to \infty$.

PROPOSITION 2.6. $\varphi_n \to \varphi$ on $\mathscr{R}$.

Proof. Let $x \in \mathscr{R}$. Without loss of generality, x is assumed real. Define $x_m, x^m \in \mathscr{C}$ as in Lemma 2.5. Then

$$\varphi x_m \leqq \varphi x \leqq \varphi x^m,$$

$$\varphi_n x_m \leqq \varphi_n x \leqq \varphi_n x^m,$$

$$\varphi_n x - \varphi x \leqq (\varphi_n x^m - \varphi x^m) + (\varphi x^m - \varphi x_m),$$

$$\varphi_n x - \varphi x \geqq (\varphi_n x_m - \varphi x_m) + (\varphi x_m - \varphi x^m).$$

Let $n \to \infty$ and then $m \to \infty$ to obtain $\varphi_n x \to \varphi x$. Thus $\varphi_n \to \varphi$ on $\mathscr{R}$.

The foregoing inequalities together with quadrature error formulas for continuous functions yield quadrature error formulas for Riemann integrable functions. Ordinarily the convergence $\varphi_n x \to \varphi x$ is slow when x is not continuous.

Proposition 2.6, with a similar proof, is due to Fejér and Stekloff; cf. Szegö [69].

25

2.8 REGULAR SETS

The pointwise convergence $\varphi_n \to \varphi$ is uniform on each totally bounded set in $\mathscr{R}$. However, this will not be sufficient for our purposes. In order to obtain a stronger result, the following concept based on Lemma 2.5 is introduced. It is in the nature of a uniform integrability condition.

DEFINITION. A set $\mathscr{F}$ of complex functions $x(t)$, $0 \leq t \leq 1$, is *regular* if for each $x \in \mathscr{F}$ and each $m = 1, 2, \ldots$ there exist $x_m, x^m \in \mathscr{C}$ such that

$$x_m \leq x \leq x^m,$$

$$\varphi(x^m - x_m) \to 0 \quad \text{uniformly for} \quad x \in \mathscr{F} \quad \text{as} \quad m \to \infty,$$

and, for each fixed m, the sets

$$\mathscr{F}_m = \{x_m : x \in \mathscr{F}\}, \qquad \mathscr{F}^m = \{x^m : x \in \mathscr{F}\}$$

are totally bounded (bounded and equicontinuous).

In view of Lemma 2.5, every regular set is in $\mathscr{R}$. A number of equivalent definitions of a regular set can be given. The condition $x_m, x^m \in \mathscr{C}$ can be replaced by $x_m, x^m \in \mathscr{R}$. We can assume that $\mathscr{F}_m$ and $\mathscr{F}^m$ are finite sets. The last sentence in Lemma 2.5 carries over to the present situation. A set $\mathscr{F} \subset \mathscr{R}$ is regular iff the sets

$$\operatorname{Re} \mathscr{F} = \{\operatorname{Re} x : x \in \mathscr{F}\}, \qquad \operatorname{Im} \mathscr{F} = \{\operatorname{Im} x : x \in \mathscr{F}\}$$

are regular. If $\mathscr{F}$ is regular, then the set

$$|\mathscr{F}| = \{|x| : x \in \mathscr{F}\}$$

is regular. Other properties and some examples of regular sets will be given shortly. First, the theorem which motivated the concept is presented.

THEOREM 2.7. $\varphi_n \to \varphi$ uniformly on each regular set $\mathscr{F} \subset \mathscr{R}$.

26

Proof. Let $x \in \mathscr{F}$. Choose x_m and x^m as in the definition of a regular set. The inequalities which appear in the proof of Proposition 2.6 hold for $x \in \mathscr{F}$ and $m \geq 1$. Since $\varphi_n \to \varphi$ uniformly on the totally bounded sets $\mathscr{F}_m$ and $\mathscr{F}^m$, and since $\varphi(x^m - x_m) \to 0$ as $m \to \infty$, uniformly for $x \in \mathscr{F}$, routine arguments imply that $\varphi_n \to \varphi$ uniformly for $x \in \mathscr{F}$.

The next proposition shows that Theorem 2.7 is a proper generalization of the fact that $\varphi_n \to \varphi$ uniformly on each totally bounded set in $\mathscr{R}$.

PROPOSITION 2.8. Every totally bounded set in $\mathscr{R}$ is regular, but not conversely.

Proof. The first assertion can be obtained directly from Lemma 2.5, or we can let $x_m = x^m = x$ in the equivalent definition of a regular set with $x_m, x^m \in \mathscr{R}$. For the converse, let $\mathscr{F}$ be the set of all characteristic functions of intervals in $[0, 1]$. Then $\mathscr{F}$ is regular, for it can be approximated in the sense of the definition by "trapezoidal" functions. However, $\mathscr{F}$ is not totally bounded since $\|x - y\| = 1$ for different $x, y \in \mathscr{F}$.

PROPOSITION 2.9. Every regular set is bounded, but not conversely.

Proof. If $\mathscr{F}$ is regular, then the approximating sets $\mathscr{F}_m$ and $\mathscr{F}^m$ are necessarily bounded, so that $\mathscr{F}$ is bounded. Now let $\mathscr{F} = \{x_n : n = 1, 2, \ldots\}$ with $x_n(t) = \cos 2\pi n t$. Clearly, $\mathscr{F}$ is bounded. For the rectangular quadrature rule,

$$\varphi_n x = \frac{1}{n} \sum_{j=1}^{n} x\left(\frac{j}{n}\right),$$

we have $\varphi_n x_n = 1$ and $\varphi x_n = 0$ for all n. By Theorem 2.7, $\mathscr{F}$ is not regular.

Regular sets have many properties in common with totally bounded sets. Subsets and closures of regular sets are regular. The convex hull of a regular set is regular. If $\mathscr{F}$ and $\mathscr{G}$ are regular, then the sets $\mathscr{F} \cup \mathscr{G}$,

$$\mathscr{F} + \mathscr{G} = \{x + y \colon x \in \mathscr{F}, \quad y \in \mathscr{G}\},$$

$$\mathscr{F}\mathscr{G} = \{xy \colon x \in \mathscr{F}, \quad y \in \mathscr{G}\}$$

are regular. Any finite linear combination of regular sets is regular.

These properties can be used to construct examples of regular sets. Let $\mathscr{F}$ be a bounded set of step functions with the number of jumps bounded uniformly. Then $\mathscr{F}$ is regular. Hence, if $\mathscr{G}$ is any bounded equicontinuous set, then $\mathscr{F} + \mathscr{G}$ is a regular set of piecewise continuous functions. There are similar but more complicated examples of regular sets of functions having infinitely many small discontinuities.

Let $\mathscr{R}_1$ consist of the same functions as $\mathscr{R}$ but topologize $\mathscr{R}_1$ by means of the seminorm $\|x\|_1$. It is rather easy to show [8] that every regular set is totally bounded in $\mathscr{R}_1$ but not conversely.

The concept of a regular set and many of its properties extend to positive linear functionals and operators defined on ordered topological vector spaces. Results include generalizations of Korovkin's theorem concerning the convergence of positive operators on $\mathscr{C}[0, 1]$. See Reference [11] by Anselone and especially [45] by James.

2.9   INTEGRAL OPERATORS WITH DISCONTINUOUS KERNELS

Consider an integral operator K on $\mathscr{R}$:

$$(Kx)(s) = \int_0^1 k(s,t)x(t)\,dt.$$

Several hypotheses on the kernel k will be examined. As before, define functions k_s and k^t by

$$k_s(t) = k^t(s) = k(s,t), \qquad 0 \le s,t \le 1.$$

The next two propositions contain modifications of some rather well-known results.

PROPOSITION 2.10. For $0 \le s \le 1$ let $k_s \in \mathscr{R}_1$ and assume

$$\|k_s - k_{s'}\|_1 \to 0 \quad \text{as} \quad s' \to s.$$

Then the foregoing convergence is uniform for $0 \leq s \leq 1$,

$$K \in [\mathscr{R}], \qquad K\mathscr{R} \subset \mathscr{C}, \qquad K \text{ is compact,}$$

$$\|K\| = \max_{0 \leq s \leq 1} \|k_s\|_1 = \max_{0 \leq s \leq 1} \int_0^1 |k(s,t)|\, dt.$$

If k is bounded, then

$$\|K\| \leq \|k\| = \sup_{0 \leq s,t \leq 1} |k(s,t)|.$$

Proof. Define $f: [0,1] \to \mathscr{R}_1$ by $f(s) = k_s$. Then f is a continuous function on a compact set. So f is uniformly continuous, which yields the first assertion, and $\|f(s)\|_1 = \|k_s\|_1$ attains a maximum. The inequalities

$$|(Kx)(s)| \leq \|k_s\|_1 \|x\|,$$
$$|(Kx)(s) - (Kx)(s')| \leq \|k_s - k_{s'}\|_1 \|x\|$$

imply that $\{Kx: \|x\| \leq 1\}$ is bounded and equicontinuous. Hence, $K \in [\mathscr{R}]$, $K\mathscr{R} \subset \mathscr{C}$, and K is compact. A simple construction gives the expression for $\|K\|$. Clearly, $\|K\| \leq \|k\|$ when k is bounded.

In particular, Proposition 2.10 pertains to any integral operator K on $\mathscr{R}$ with a continuous kernel k. The next result indicates that it also applies if k is a suitable limit of a sequence of continuous kernels k^m, $m = 1, 2, \ldots$.

PROPOSITION 2.11. Let $k_s \in \mathscr{R}_1$ for $0 \leq s \leq 1$. For each (superscript) $m = 1, 2, \ldots$, let K^m be an integral operator with a continuous kernel k^m. Let

$$\|k_s^m - k_s\|_1 \to 0 \quad \text{uniformly for} \quad 0 \leq s \leq 1 \quad \text{as} \quad m \to \infty,$$

where $k_s^m(t) = k^m(s,t)$. Then k and K satisfy the hypotheses and conclusions of Proposition 2.10. Moreover,

$$\|K^m - K\| \to 0.$$

Proof. Define $f, f^m: [0,1] \to \mathscr{R}_1$ by $f(s) = k_s$ and $f^m(s) = k_s^m$. Then each f^m is continuous and $f^m(s) \to f(s)$ uniformly. So f is continuous and Proposition 2.10 applies to k and K. Finally,

$$\|K^m - K\| = \max_{0 \leq s \leq 1} \|k_s^m - k_s\|_1 \to 0.$$

(The conclusions from Proposition 2.10 that $K\mathscr{R} \subset \mathscr{C}$ and K is compact also follow from $K^m \mathscr{R} \subset \mathscr{C}$ and $\|K^m - K\| \to 0$.)

In the following proposition, k is the limit of a pair of sequences $\{k_m: m = 1, 2, \ldots\}$ and $\{k^m: m = 1, 2, \ldots\}$ of continuous kernels that bracket k. The notation

$$k_{ms}(t) = k_m(s,t), \qquad k_s^m(t) = k^m(s,t)$$

is used.

PROPOSITION 2.12. For $m = 1, 2, \ldots$ let k_m and k^m be continuous kernels such that

$$k_m \le k \le k^m,$$

$\|k_s^m - k_{ms}\|_1 \to 0$ uniformly for $0 \le s \le 1$ as $m \to \infty$.

Then the hypotheses and conclusions of Propositions 2.10 and 2.11 are satisfied. Moreover, k is bounded and $\{k_s: 0 \le s \le 1\}$ is regular.

Proof. Clearly, k is bounded and $\{k_s: 0 \le s \le 1\}$ is regular. Hence, $k_s \in \mathscr{R}_1$ and

$$\|k_s^m - k_s\|_1 \le \|k_s^m - k_{ms}\|_1, \qquad 0 \le s \le 1.$$

So Propositions 2.11 and 2.10 apply in turn.

2.10 OPERATOR APPROXIMATIONS

Throughout this section let K be an integral operator on $\mathscr{R}$ with a kernel k such that

(a) $\{k_s: 0 \le s \le 1\}$ is regular,
(b) $\|k_s - k_{s'}\|_1 \to 0$ as $s' \to s$, $0 \le s, s' \le 1$,
(c) $k^t \in \mathscr{R}$, $0 \le t \le 1$.

Note that (a) implies each $k_s \in \mathscr{R}_1$ and k is bounded.

Any continuous kernel satisfies (a)–(c). So does any Volterra kernel which is continuous for $0 \le t \le s \le 1$ and zero for $0 \le s < t \le 1$. This is easily verified with the aid of Proposition 2.12. A more general example is a bounded, piecewise continuous kernel k with the discontinuities confined to a finite number of continuous curves $t = t_i(s)$, $i = 1, 2, \ldots, m$. Such a "mildly discontinuous" kernel satisfies the hypotheses of Proposition 2.12.

Define operators K_n, $n = 1, 2, \ldots$, on $\mathscr{R}$ by

$$(K_n x)(s) = \sum_{j=1}^{n} w_{nj} k(s, t_{nj}) x(t_{nj}),$$

where, as in §2.7 and §2.8, the quadrature formula satisfies

(d) $\varphi_n \to \varphi$ on $\mathscr{R}$,

(e) $w_{nj} \geqq 0$, $1 \leqq j \leqq n$.

By (c), $K_n \mathscr{R} \subset \mathscr{R}$. Since k is bounded, each $K_n \in [\mathscr{R}]$ and

$$\|K_n\| = \sup_{0 \leqq s \leqq 1} \sum_{j=1}^{n} w_{nj} |k(s, t_{nj})| \leqq b \|k\|.$$

Thus, $\{K_n\}$ is bounded. Each K_n is compact because $K_n \mathscr{R}$ is finite dimensional.

THEOREM 2.13. Assume (a)–(e). Then

$$K, K_n \in [\mathscr{R}], \qquad K\mathscr{R} \subset \mathscr{C}, \qquad K \text{ is compact},$$

$\{K_n\}$ is collectively compact,

$K_n \to K$,

$\|K_n\| \to \|K\|$.

Proof. We have shown already that each $K_n \in [\mathscr{R}]$. Proposition 2.10 yields $K \in [\mathscr{R}]$, $K\mathscr{R} \subset \mathscr{C}$, and K compact. The equations

$$(Kx)(s) = \varphi(k_s x), \qquad (K_n x)(s) = \varphi_n(k_s x),$$

$$\|K\| = \max_{0 \leqq s \leqq 1} \varphi(|k_s|), \qquad \|K_n\| = \sup_{0 \leqq s \leqq 1} \varphi_n(|k_s|),$$

together with (a) and Theorem 2.7, imply $K_n \to K$ and $\|K_n\| \to \|K\|$. Note that

$$|(K_n x)(s) - (K_n x)(s')| \leqq \varphi_n(|k_s - k_{s'}|) \|x\|,$$

where, again by Theorem 2.7,

$$\varphi_n(|k_s - k_{s'}|) \to \varphi(|k_s - k_{s'}|) = \|k_s - k_{s'}\|_1$$

uniformly for $0 \leqq s,s' \leqq 1$. In view of (b) and Proposition 2.10,

$$(K_n x)(s) - (K_n x)(s') \to 0 \quad \text{as} \quad n \to \infty \quad \text{and} \quad s' \to s,$$

$$\text{uniformly for} \quad \|x\| \leqq 1 \quad \text{and} \quad 0 \leqq s \leqq 1.$$

This implies that the sequence of sets $\{K_n x: \|x\| \leqq 1\}$ is asymptotically equicontinuous. These sets are bounded uniformly and each of them is totally bounded, because $\{K_n\}$ is bounded and each K_n is compact. By Proposition 2.4, the set $\{K_n x: n \geqq 1, \|x\| \leqq 1\}$ is totally bounded. Thus, $\{K_n\}$ is collectively compact.

The conclusion of Theorem 2.13 will be obtained in another way for a more restrictive class of kernels in Chapter 3, §3.2.

Since $K_n \to K$ and $\{K_n\}$ is collectively compact, the approximation theory of Chapter 1 applies to the equations in $\mathscr{R}$,

$$(I - K)x = y, \qquad (I - K_n)x_n = y.$$

Suppose $(I - K)^{-1}$ and $(I - K_n)^{-1}$ exist. The evaluation of x_n and the estimation of $\|x_n - x\|$ can be accomplished in the same way as was done for continuous kernels in §2.4 and §2.5. The convergence $\|x_n - x\| \to 0$ is generally slow because of the inefficiency of the numerical integration of discontinuous functions. Better methods for handling particular types of operators are described in the next chapter.

In conclusion, note that if $y \in \mathscr{C}$, then $x = y + Kx \in \mathscr{C}$ since $K\mathscr{R} \subset \mathscr{C}$. But ordinarily $x_n \notin \mathscr{C}$ when k is discontinuous, in which case the continuous function x is being approximated by discontinuous functions x_n.

chapter 3

VARIANTS OF THE THEORY
AND FURTHER APPLICATIONS

3.1 INTRODUCTION

Various special techniques designed for particular classes of problems are presented in the first six sections of this chapter. Sections 3.7 and 3.10 present modifications of the theory, Section 3.8 concerns the approximate solution of transport equations, and applications to boundary value problems for partial differential equations are mentioned briefly in Section 3.9. Each topic is a synopsis of a more complete development which has appeared recently in the mathematical literature.

The analysis takes place in an arbitrary Banach space $\mathscr{X}$ and in the spaces $\mathscr{C} = \mathscr{C}[0, 1]$ and $\mathscr{R} = \mathscr{R}[0, 1]$ of Chapter 2.

3.2 PERTURBATIONS OF
COLLECTIVELY COMPACT OPERATOR SEQUENCES

We shall show that certain limits of operators satisfying the hypotheses of the theory in Chapter 1 also satisfy the same hypotheses. An application

33

is to integral operators which are limits in a particular sense of integral operators with continuous kernels. This material is taken from Reference [8] by Anselone.

Let $K, K_n, K^m, K_n^m \in [\mathscr{X}]$ for $m, n = 1, 2, \ldots$ where m is a superscript. Recall that if each K^m is compact and $\|K^m - K\| \to 0$, then K is compact. Analogously, if $\{K_n^m : n \geq 1\}$ is collectively compact for each m, and if $\|K_n^m - K_n\| \to 0$ uniformly in n as $m \to \infty$, then $\{K_n : n \geq 1\}$ is collectively compact. The following more complicated result is similar in nature.

PROPOSITION 3.1. Assume K_n is compact for each n, $\{K_n^m : n \geq 1\}$ is collectively compact for each m, and

$$\varlimsup_{m \to \infty} \lim_{n \to \infty} \|K_n^m - K_n\| = 0.$$

Then $\{K_n : n \geq 1\}$ is collectively compact.

Proof. For each $\varepsilon > 0$ there exist $M = M(\varepsilon)$ and $N = N(\varepsilon)$ such that

$$\|K_n^M x - K_n x\| \leq \|K_n^M - K_n\| < \varepsilon \quad \text{for} \quad n \geq N, \qquad x \in \mathscr{B},$$

where $\mathscr{B}$ is the closed unit ball in $\mathscr{X}$. Therefore, $\{K_n : n \geq N\}\mathscr{B}$ has the ε-net $\{K_n^M : n \geq N\}\mathscr{B}$ which, by the second hypothesis, is totally bounded. Since each K_n is compact, $\{K_n : n < N\}\mathscr{B}$ has a finite ε-net. Hence, $\{K_n : n \geq 1\}\mathscr{B}$ has a totally bounded ε-net for each $\varepsilon > 0$, which implies that it is totally bounded. So $\{K_n : n \geq 1\}$ is collectively compact.

PROPOSITION 3.2. Assume K_n is compact for each n, $\{K_n^m : n \geq 1\}$ is collectively compact for each m, and

$$\|K_n^m - K_n\| \to \|K^m - K\| \quad \text{as} \quad n \to \infty, \qquad m \geq 1,$$

$$\|K^m - K\| \to 0 \qquad\qquad \text{as} \quad m \to \infty,$$

$$K_n^m \to K^m \qquad\qquad\qquad \text{as} \quad n \to \infty, \qquad m \geq 1.$$

Then $K_n \to K$, $\{K_n : n \geq 1\}$ is collectively compact and, hence, K is compact.

Proof. By the triangle inequality,

$$\|K_n x - K x\| \leq \|K_n x - K_n^m x\| + K_n^m x - K^m x\| + \|K^m x - K x\|.$$

Let $n \to \infty$ and then $m \to \infty$ to show that $K_n \to K$. Reference to Proposition 3.1 completes the proof.

For an application of the preceding results, consider an integral operator $K \in [\mathscr{R}]$,

$$(Kx)(s) = \int_0^1 k(s,t)x(t)\, dt,$$

where k satisfies the conditions (a)–(c) of Chapter 2, §2.10. In addition, assume (cf. Propositions 2.11 and 2.12) there exist operators $K^m \in [\mathscr{R}]$, $m \geq 1$, with continuous kernels k^m such that $\|K^m - K\| \to 0$.

Define K_n, $K_n^m \in [\mathscr{R}]$ for $m,n \geq 1$ by

$$(K_n x)(s) = \sum_{j=1}^n w_{nj} k(s,t_{nj}) x(t_{nj}),$$

$$(K_n^m x)(s) = \sum_{j=1}^n w_{nj} k^m(s,t_{nj}) x(t_{nj}),$$

where, once again, the quadrature formula satisfies

$$\sum_{j=1}^n w_{nj} x(t_{nj}) \to \int_0^1 x(t)\, dt, \qquad x \in \mathscr{R},$$

$$w_{nj} \geq 0, \qquad 1 \leq j \leq n.$$

Clearly, each K_n is compact. By Theorem 2.13, $K_n \to K$ and $\|K_n\| \to \|K\|$.

In the present circumstances, another conclusion of Theorem 2.13, namely that $\{K_n\}$ is collectively compact, can be obtained alternatively as follows. For each m, $K_n^m \to K^m$ as $n \to \infty$ and $\{K_n^m: n \geq 1\}$ is collectively compact by Propositions 2.1 and 2.2. Define $(K^m - K)_n = K_n^m - K_n$ and replace K by $K^m - K$ in $\|K_n\| \to \|K\|$ to obtain

$$\|K_n^m - K_n\| \xrightarrow[n]{} \|K^m - K\| \xrightarrow[m]{} 0.$$

Then $\{K_n\}$ is collectively compact by Proposition 3.1 or 3.2. The inequality in the proof of the latter provides a possibly useful error bound. It gives a means of taking advantage of the error analysis for integral operators with continuous kernels.

3.3 FACTORIZATION OF THE KERNEL

This material is adapted from Atkinson [19].

Often the kernel of an integral operator K can be factored, $k(s,t) = r(s,t)\sigma(s,t)$, with $r(s,t)$ continuous or smoother and with $\sigma(s, t)$ a relatively simple function which may be discontinuous or even singular. Then it may be possible to obtain good numerical integration approximations for K by approximating only $r(s,t)$ while integrating $\sigma(s,t)$ exactly. If so, the effect of any lack of smoothness of $k(s,t)$ on the accuracy of approximation is largely mitigated.

Let K be an integral operator,

$$(Kx)(s) = \int_0^1 k(s,t)x(t)\, dt.$$

Assume that the functions $k_s(t) = k(s,t)$ satisfy

$$k_s \in \mathcal{L}^1(0,1), \qquad 0 \le s \le 1,$$

$$\|k_s - k_{s'}\|_1 \to 0 \quad \text{as} \quad s' \to s.$$

As in Chapter 2, §2.9, the foregoing convergence is necessarily uniform, $K \in [\mathscr{C}]$, K is compact, and $\|K\| = \max \|k_s\|_1$.

Suppose that $k(s,t)$ is factored,

$$k(s,t) = r(s,t)\, \sigma(s,t),$$

where $r(s,t)$ is continuous for $0 \le s,t \le 1$ and the functions $\sigma_s(t) = \sigma(s,t)$ satisfy

$$\sigma_s \in \mathcal{L}^1(0,1), \qquad 0 \le s \le 1,$$

$$\|\sigma_s - \sigma_{s'}\|_1 \to 0 \quad \text{as} \quad s' \to s.$$

These conditions on $r(s,t)$ and $\sigma(s,t)$ imply the previously stated properties of $k(s,t)$. Now

$$(Kx)(s) = \int_0^1 [r(s,t)x(t)]\sigma(s,t)\, dt,$$

where $r(s,t)x(t)$ is continuous.

Operators $K_n \in [\mathscr{C}]$, $n = 1, 2, \ldots$, may be defined by numerical integration of $r(s,t)x(t)$, with σ as a weight function. Various quadrature rules can be used for this purpose. A rather general procedure will be described. Thus, let A_n, $n = 1, 2, \ldots$, be continuous, not necessarily linear, functions on $\mathscr{C}$ into $\mathscr{C}$ such that $A_n \to I$, i.e.,

$$A_n x \to x, \qquad x \in \mathscr{C}.$$

For example, A_n could represent interpolation, Tchebycheff or spline approximation, or series truncation. Define operators $K_n \in [\mathscr{C}]$, $n = 1, 2, \ldots$, by

$$(K_n x)(s) = \int_0^1 \{A_n[r(s,t)x(t)]\}\sigma(s,t)\, dt,$$

where A_n acts with respect to t.

If $A_n \in [\mathscr{C}]$ for all n, then it follows from $A_n x \to x$ on $\mathscr{C}$ and the Banach-Steinhaus theorem that $\{A_n\}$ is equicontinuous and, hence, that $A_n x \to x$ uniformly on each bounded equicontinuous subset of $\mathscr{C}$. Assume the last condition. Then

$$A_n[r(s,t)x(t)] \to r(s,t)x(t) \quad \text{uniformly in} \quad s \quad \text{and} \quad t,$$

$$\{A_n[r(s,t)x(t)]: n \geq 1\} \quad \text{is bounded and equicontinuous.}$$

Standard arguments yield

$$\{K_n\} \text{ is collectively compact,}$$

$$K_n \to K,$$

$$\|K_n x - Kx\| \leq \max_{0 \leq s \leq 1} \|(A_n - I)[r(s,t)x(t)]\| \, \|\sigma_s\|_1.$$

Thus, the general approximation theory of Chapter 1 is applicable.

As a particular case, let A_n correspond to piecewise linear interpolation with subdivision points $t_{nj} = j/n$, $j = 0, 1, \ldots, n$. Then $(K_n x)(s)$ reduces to

$$(K_n x)(s) = \sum_{j=0}^{n} w_{nj}(s)r(s,t_{nj})x(t_{nj}),$$

where

$$w_{nj}(s) = \int_{(j-1)/n}^{(j+1)/n} (1 - |nt - j|)\sigma(s,t)\, dt$$

with $\sigma(s,t) = 0$ for $t < 0$ and $t > 1$ to make $w_{n0}(s)$ and $w_{nn}(s)$ correct. In order to evaluate $(K_n x)(s)$ explicity, it is necessary to be able to integrate $\sigma(s,t)$ and $t\sigma(s,t)$ with respect to t in closed form. If A_n represents piecewise polynomial interpolation, then $(K_n x)(s)$ has the same general form as above, with $w_{nj}(s)$ expressed in terms of integrals involving $t^m \sigma(s,t)$ for various m. These integrals can be evaluated if $\sigma(s,t)$ is simple enough, e.g., $\sigma(s,t) = \ln|s - t|$ or $\sigma(s,t) = |s - t|^\alpha$ with $\alpha > -1$. When $\sigma(s,t) \equiv 1$, standard numerical integration rules are obtained and the analysis in Chapter 2 applies.

An element of choice is involved in the factorization of a kernel $k(s,t)$. It is clearly advantageous for $r(s,t)$ to be as smooth as possible, while $\sigma(s,t)$ must be simple enough to enable $(K_n x)(s)$ to be evaluated. Thus, a factor $|s - t|^{\frac{1}{2}}$, even though continuous, should be put into $\sigma(s,t)$ rather than into $r(s,t)$.

An immediate extension of the method of this section pertains to a kernel $k(s, t)$ which can be expressed in the form

$$k(s,t) = \sum_{i=1}^{m} r_i(s,t)\sigma_i(s,t),$$

where the functions $r_i(s,t)$ and $\sigma_i(s,t)$ have the properties imposed above on $r(s, t)$ and $\sigma(s,t)$. For example, let $k(s,t) = Y_0(|s - t|)$, where Y_0 is the Bessel function of the second kind and zero order. Since $Y_0(z) = f(z)\ln z + g(z)$, with f and g analytic, $k(s,t)$ can be expressed in the desired form. Another interesting example is

$$k(s,t) = \ln|\cos s - \cos t|.$$

The factorization with

$$r(s,t) = |s - t|^{\frac{1}{2}}\ln|\cos s - \cos t|, \qquad \sigma(s,t) = |s - t|^{-\frac{1}{2}}$$

has the disadvantage that $\partial r/\partial t$ is unbounded, which results in numerical inefficiency. On the other hand, trigonometric identities yield a superior representation,

$$k(s,t) = \sum_{i=1}^{4} r_i(s,t)\sigma_i(s,t),$$

where

$$r_1(s,t) = \ln\left[\frac{\sin\left(\dfrac{t - s}{2}\right)\sin\left(\dfrac{t + s}{2}\right)}{\left(\dfrac{t - s}{2}\right)(t + s)(2\pi - t - s)}\right],$$

$$r_2(s,t) = r_3(s,t) = r_4(s,t) = 1,$$
$$\sigma_1(s,t) = 1, \qquad\qquad\qquad \sigma_2(s,t) = \ln|s - t|,$$
$$\sigma_3(s,t) = \ln(2\pi - s - t), \qquad \sigma_4(s,t) = \ln(s + t).$$

Numerical results for this kernel are given in Appendix 2 and in [19].

3.4 VOLTERRA AND RELATED KERNELS

This method, adapted from Davis and Jacoby [34], pertains to Volterra operators and other operators with kernels $k(s,t)$ which are discontinuous or which have discontinuous derivatives on the diagonal $s = t$.

To begin with, consider a Volterra operator on $\mathscr{C}$,

$$(Kx)(s) = \int_0^s k(s,t)x(t)\,dt,$$

where $k(s,t)$ is continuous or smoother for $0 \le t \le s \le 1$. Express $k(s,t)$

as a finite Taylor expansion plus remainder:

$$k(s,t) = k_1(s,t) + k_2(s,t), \qquad 0 \leq t \leq s \leq 1,$$

where, for some $m \geq 0$,

$$k_1(s,t) = \sum_{l=0}^{m} a_l(s)(s-t)^l, \qquad a_l(s) = \frac{(-1)^l}{l!} \frac{\partial^l k(s,t)}{\partial t^l} \quad \text{at} \quad t = s.$$

In particular,

$$k_1(s,s) = k(s,s), \qquad k_2(s,s) = 0.$$

Define $k_2(s,t) = 0$ for $0 \leq s < t \leq 1$. Then $k_2(s,t)$ is continuous on the unit square.

Express K as a sum,

$$K = K_1 + K_2,$$

where K_1 and K_2 are defined by

$$(K_1 x)(s) = \int_0^s k_1(s,t)x(t)\, dt = \sum_{l=0}^{m} a_l(s) \int_0^s (s-t)^l x(t)\, dt,$$

$$(K_2 x)(s) = \int_0^1 k_2(s,t)x(t)\, dt.$$

The analysis in the first part of Chapter 2 applies to the numerical integration approximation of K_2. The efficiency of the approximation depends on the smoothness of k and on the size of m.

It remains to consider the Volterra integral operators

$$\int_0^s (s-t)^l x(t)\, dt.$$

As in the preceding section, we introduce approximations of the form

$$\int_0^s (s-t)^l (A_n x)(t)\, dt,$$

where each A_n maps $\mathscr{C}$ into $\mathscr{C}$ and $A_n x \to x$ for $x \in \mathscr{C}$, uniformly on bounded equicontinuous subsets. Then, putting all the pieces together, we obtain a sequence of operators $K_n \in [\mathscr{C}]$ such that

$$\{K_n\} \text{ is collectively compact,}$$

$$K_n \to K.$$

Thus, the general theory applies.

39

For example, let A_n represent piecewise linear interpolation with subdivision points $t_{nj} = j/n$, $j = 0, 1, \ldots, n$. Then elementary but rather tedious analysis yields

$$\int_0^s (s - t)^l (A_n x)(t)\, dt = \sum_{j=0}^n w_{nj}(s) x(t_{nj}),$$

where

$$w_{n0}(s) = \frac{s^{l+1}}{l+1} - \Phi_{n0}(s) + \Phi_{n1}(s),$$

$$w_{nj}(s) = \Phi_{n,j-1}(s) - 2\Phi_{nj}(s) + \Phi_{n,j+1}(s), \qquad 1 \leq j \leq n,$$

with

$$\Phi_{nj}(s) = \max \left\{ 0, \ \frac{\left(n\left(s - \dfrac{j}{n} \right) \right)^{l+2}}{(l+1)(l+2)} \right\}.$$

Similar but more complicated formulas can be derived when A_n corresponds to higher order polynomial interpolation.

Methods of approximation slightly different from the foregoing were given by Davis and Jacoby [34]. Their analogue of the trapezoidal rule is described as follows: For $j/n \leq s < (j+1)/n$ consider

$$\int_0^s (s - t)^l x(t)\, dt = \int_0^{j/n} (s - t)^l x(t)\, dt + \int_{j/n}^s (s - t)^l x(t)\, dt.$$

The first integral on the right is approximated by means of the ordinary trapezoidal rule with subdivision points i/n, $i = 0, \ldots, j$. In the second, $x(t)$ is replaced by the linear interpolant,

$$x\left(\frac{j}{n} \right)(j + 1 - ns) + x\left(\frac{j+1}{n} \right)(ns - j),$$

and then the integration is performed. Special Simpson's and higher order quadrature rules are derived in a similar fashion. See [34] for a detailed remainder analysis.

The approach of Davis and Jacoby has both advantages and disadvantages as compared to the approach given earlier in this section. On the one hand, it does involve simpler formulas and, hence, less computation for each step in the procedure. On the other hand, the resulting approximation of the Volterra integral operator with kernel $(s - t)^l$ does not map $\mathscr{C}$ into $\mathscr{C}$; the image functions usually have small discontinuities at $s = j/n, j = 1, \ldots,$ $n - 1$. This complicates the theoretical analysis.

Consider now a Fredholm integral operator on $\mathscr{C}$,

$$(Kx)(s) = \int_0^1 k(s,t)x(t)\,dt,$$

where $k(s,t)$ has m continuous partial derivatives on each of the two triangles $0 \leq t \leq s \leq 1$ and $0 \leq s \leq t \leq 1$. Thus, simple discontinuities of the derivatives may occur on the diagonal $s = t$. Such integral operators arise from boundary value problems. Let

$$g_l(s) = \frac{\partial^l}{\partial t^l}\, k(s,s-) - \frac{\partial^l}{\partial t^l}\, k(s,s+), \qquad l = 0, 1, \ldots, m.$$

Then

$$K = K_1 + K_2,$$

where

$$(K_2 x)(s) = \sum_{l=0}^{m} \frac{(-1)^l}{l!}\, g_l(s) \int_0^s (s - t)^l x(t)\,dt$$

and K_1 has a kernel with m continuous partial derivatives in the unit square.

One example, treated above, is a Volterra kernel. Another is the one dimensional Green's function

$$k(s,t) = \begin{cases} t(1 - s), & 0 \leq t \leq s \leq 1, \\ s(1 - t), & 0 \leq s \leq t \leq 1, \end{cases}$$

associated with a two-point boundary value problem.

For further details see Appendix 2 and Davis and Jacoby [34].

3.5 INTEGRAL EQUATIONS WITHOUT UNIQUE SOLUTIONS

The material in this section is a condensation with alterations of a paper by Atkinson [20].

Once again, let K be an integral operator on $\mathscr{C}$,

$$(Kx)(s) = \int_0^1 k(s,t)x(t)\,dt.$$

Although it is not essential for what follows, suppose that k is continuous on the closed unit square. Assume also that $(I - K)^{-1}$ does not exist. Thus, for each $y \in (I - K)\mathscr{C}$, the equation $(I - K)x = y$ has a nonunique solution.

Particular solutions of such equations frequently are of interest and have to be obtained by approximation procedures. The nonuniqueness causes difficulties with the direct use of numerical integration. To overcome these difficulties, $I - K$ will be replaced by a related invertible operator.

By the Fredholm alternative (cf. Appendix 1), the null space and range of $I - K$ satisfy

$$\dim \mathcal{N}(I - K) = \operatorname{codim} (I - K)\mathscr{C} = m$$

for some positive integer m. Thus, there exists a subspace $\mathcal{M}$ of $\mathscr{C}$ such that

$$\mathscr{C} = (I - K)\mathscr{C} \oplus \mathcal{M}, \qquad \dim \mathcal{M} = m.$$

Choose bases $\{\varphi_1, \ldots, \varphi_m\}$ for $\mathcal{N}(I - K)$ and $\{\psi_1, \ldots, \psi_m\}$ for $\mathcal{M}$. Define $L \in [\mathscr{C}]$ by

$$Lx = \sum_{i=1}^{m} x(p_i)\psi_i$$

where each $p_i \in [0, 1]$. Then $L\mathscr{C} \subset \mathcal{M}$ and L is compact. The restriction of L to $\mathcal{N}(I - K)$ has the matrix representation $[\varphi_i(p_j)]$ with respect to $\{\varphi_1, \ldots, \varphi_m\}$ and $\{\psi_1, \ldots, \psi_m\}$. Henceforth, assume that $\det [\varphi_i(p_j)] \neq 0$. Equivalently, L maps $\mathcal{N}(I - K)$ one-to-one onto $\mathcal{M}$, or

$$\mathcal{N}(I - K) \cap \mathcal{N}(L) = \Theta = \{0\}.$$

If $(I - K - L)x = 0$, then

$$(I - K)x = Lx \in (I - K)\mathscr{C} \cap \mathcal{M} = \Theta,$$

$$(I - K)x = 0 = Lx, \qquad x = 0.$$

Thus $(I - K - L)^{-1}$ exists. Since $K + L$ is compact,

$$(I - K - L)^{-1} \in [\mathscr{C}].$$

If $(I - K - L)x = y$ and $y \in (I - K)\mathscr{C}$, then $\mathscr{C} = (I - K)\mathscr{C} \oplus \mathcal{M}$ implies that $(I - K)x = y$ and $Lx = 0$. Thus, the restriction of $(I - K - L)^{-1}$ to $(I - K)\mathscr{C}$ is a *selective* inverse of $I - K$.

Approximate K by numerical integration:

$$(K_n x)(s) = \sum_{j=1}^{n} w_{nj} k(s, t_{nj}) x(t_{nj}),$$

where the quadrature formula has the properties in Chapter 2, §2.2. Then

$$K_n + L \to K + L,$$

$\{K_n + L\}$ is collectively compact,

and the approximation theory in Chapter 1 relates $(I - K - L)^{-1}$ and $(I - K_n - L)^{-1}$, $n = 1, 2, \ldots$. The equation $(I - K_n - L)x_n = y$, when written out, is

$$x_n(s) - \sum_{j=1}^{n} w_{nj} k(s, t_{nj}) x_n(t_{nj}) - \sum_{i=1}^{m} x_n(p_i) \psi_i(s) = y(s).$$

The substitution of $s = t_{nj}$, $j = 1, \ldots, n$ and $s = p_j$, $i = 1, \ldots, m$, yields an equivalent finite algebraic system. There are $m + n$ equations if the numbers p_i and t_{nj} are all distinct; otherwise there are fewer equations. Thus, at least in principle, we can determine functions $x_n \in \mathscr{C}$ which converge to a particular solution x of $(I - K)x = y$ for any given $y \in \mathscr{C}$.

To carry out the foregoing procedure, a basis $\{\psi_1, \ldots, \psi_m\}$ for a complement of $(I - K)\mathscr{C}$ is needed. For this purpose it is convenient to regard K also as an operator on the larger space $\mathscr{L}^2 = \mathscr{L}^2(0,1)$. The adjoint operator $K^* \in [\mathscr{L}^2]$ is expressed by

$$(K^* x)(s) = \int_0^1 \overline{k(t,s)} x(t) \, dt.$$

It follows from $(Ku, v) = (u, K^* v)$ that

$$[(I - K)\mathscr{L}^2]^\perp = \mathscr{N}(I - K^*),$$

$$\mathscr{L}^2 = (I - K)\mathscr{L}^2 \oplus \mathscr{N}(I - K^*).$$

Since k is continuous by assumption, $K^* \mathscr{L}^2 \subset \mathscr{C}$ and $\mathscr{N}(I - K^*) \subset \mathscr{C}$. Therefore,

$$\mathscr{C} = (I - K)\mathscr{C} \oplus \mathscr{N}(I - K^*).$$

As a later example will illustrate, theoretical analysis may yield a basis for the complement $\mathscr{N}(I - K^*)$ of $(I - K)\mathscr{C}$.

The definition of the operator L also requires the selection of points $p_i \in [0,1]$ such that $\mathscr{N}(I - K) \cap \mathscr{N}(L) = \Theta$ or, equivalently, such that $\det [\varphi_i(p_j)] \neq 0$ for some basis $\{\varphi_1, \ldots, \varphi_m\}$ of $\mathscr{N}(I - K)$. This is not a severe problem; a random choice of the points p_i probably suffices. Better yet, several trial selections should include one for which the norm of $(I - K_n - L)^{-1}$ is not too great. This is important for the efficient calculation of approximate solutions of $(I - K)x = y$.

The approximation method described above can be applied to the interior Neumann problem for $\Delta u = 0$ on a simply connected planar region D with a smooth boundary Γ. This problem has the integral equation formulation

$$\rho(x,y) - \frac{1}{\pi} \int_\Gamma \rho(\xi,\eta) \frac{d}{dv(x,y)} \left[\ln \frac{1}{r} \right] ds = f(x,y), \qquad (x,y) \in \Gamma,$$

where $r = [(x - \xi)^2 + (y - \eta)^2]^{\frac{1}{2}}$, s is arc length on Γ, and $v(\xi,\eta)$ is the interior normal to Γ at (ξ,η). The null space of the adjoint equation is spanned by the single function $\psi(x,y) \equiv 1$. Hence, for a suitable choice of $(x_0,y_0) \in \Gamma$, the equation

$$\rho(x,y) - \frac{1}{\pi} \int_\Gamma \rho(\xi,\eta) \frac{d}{dv(x,y)} \left[\ln \frac{1}{r} \right] ds + \rho(x_0,y_0) = f(x,y), \qquad (x,y) \in \Gamma,$$

is uniquely solvable. Atkinson [20] specializes Γ to an elliptical boundary and presents numerical results.

3.6 INTEGRAL EQUATIONS ON $[0,\infty)$

The following material is adapted from Atkinson [22]. The reader is referred to this paper for further details, more general results, and examples.

Let $\tilde{\mathscr{C}}[0,\infty)$ be the Banach space of bounded, continuous, real or complex functions on $[0,\infty)$ with the sup norm. Consider an integral operator K on $\tilde{\mathscr{C}}[0,\infty)$,

$$(Kx)(s) = \int_0^\infty k(s,t)x(t)\, dt,$$

where the functions $k_s(t) = k(s,t)$ satisfy

$$k_s \in \mathscr{L}^1(0,\infty), \qquad s \geq 0,$$

$$\sup_{s \geq 0} \|k_s\|_1 < \infty,$$

$$\|k_s - k_{s'}\|_1 \to 0 \quad \text{as} \quad s' \to s, \quad \text{uniformly for} \quad s \geq 0.$$

Then K is a bounded linear operator on $\tilde{\mathscr{C}}[0,\infty)$ and

$$\|K\| = \sup_{s \geq 0} \|k_s\|_1.$$

Let $\mathscr{C}_0[0,\infty]$ denote the subspace of $\tilde{\mathscr{C}}[0,\infty)$ consisting of the functions which vanish at ∞. The final hypothesis on the kernel,

$$\int_0^\beta |k(s,t)|\, dt \to 0 \quad \text{as} \quad s \to \infty, \qquad 0 \leq \beta < \infty,$$

implies that K maps $\mathscr{C}_0[0,\infty]$ into $\mathscr{C}_0[0,\infty]$.

Although K maps each bounded subset of $\tilde{\mathscr{C}}[0,\infty)$ into a bounded equicontinuous set, K is not necessarily compact because the Arzelà-Ascoli lemma does not carry over to $\tilde{\mathscr{C}}[0,\infty)$. However, if $\|k_s\|_1 \to 0$ as $s \to \infty$, then K is a compact operator on $\mathscr{C}_0[0,\infty]$, for the Arzelà-Ascoli lemma does extend to this space. Unfortunately, the foregoing condition fails to hold for a large and significant class of kernels which satisfy the previously stated hypotheses. In particular, there is the Picard kernel, $k(s,t) = e^{-|s-t|}$. More general examples are furnished by

$$k(s,t) = \kappa(s - t), \qquad \kappa \in \mathscr{L}^1(-\infty, \infty),$$

and

$$k(s,t) = \frac{\omega(s)}{\omega(t)} \kappa(s - t),$$

where

$$k \in \mathscr{L}^1(-\infty, \infty), \qquad \omega \in \mathscr{C}[0,\infty), \qquad \omega > 0,$$

$$\lim_{s \to \infty} (s + 1)^n e^{\alpha s} \omega(s) > 0,$$

$$u^j e^{-\alpha u} \kappa(u) \in \mathscr{L}^1(-\infty, \infty), \qquad j = 0, 1, \ldots, n,$$

for some $n \geq 0$ and some real α such that $|\alpha| + n > 0$.

The operator K with kernel k will be approximated by

$$(K_\beta x)(s) = \int_0^\beta k(s,t) x(t)\, dt, \qquad \beta \geq 0.$$

These are compact operators on $\tilde{\mathscr{C}}[0,\infty)$ into $\mathscr{C}_0[0,\infty]$. Let

$$\|x\|_\beta = \sup_{s \geq \beta} |x(s)|.$$

Then, for each $x \in \tilde{\mathscr{C}}[0,\infty)$ and each $s \geq 0$,

$$|(K_\beta x)(s) - (Kx)(s)| \leq \int_\beta^\infty |k(s,t)|\, dt\, \|x\|_\beta \leq \|K\|\, \|x\|_\beta,$$

$$(K_\beta x)(s) \to (Kx)(s) \quad \text{as} \quad \beta \to \infty.$$

The convergence is uniform for s in any finite interval since the functions

$K_\beta x$, $\beta \geqq 0$, are equicontinuous. If $x \in \mathscr{C}_0[0,\infty]$, then $\|x\|_\beta \to 0$ and $K_\beta \to K$ as $\beta \to \infty$.

If $(I - K_\beta)^{-1}$ exists and is uniformly bounded for β sufficiently large then, by Lemma 1.5,

$$(I - K_\beta)^{-1} \to (I - K)^{-1} \quad \text{on} \quad \mathscr{C}_0[0,\infty].$$

The determination of $(I - K_\beta)^{-1}$ for a fixed β essentially involves only $\mathscr{C}[0,\beta]$. So the collectively compact operator approximation theory can be applied to this task.

The analysis for $\widetilde{\mathscr{C}}[0,\infty)$ is more complicated. Suppose that $(I - K)^{-1}$ and $(I - K_\beta)^{-1}$ exist as operators on $\widetilde{\mathscr{C}}[0,\infty)$ and $(I - K_\beta)^{-1}$ is bounded uniformly for β sufficiently large. Fix $y \in \widetilde{\mathscr{C}}[0,\infty)$ and let

$$(I - K)x = y, \qquad (I - K_\beta)x_\beta = y.$$

Then $x_\beta(s) \to x(s)$ as $\beta \to \infty$ uniformly for s in any finite interval. We shall sketch a proof of this assertion. Let $e_\beta = x - x_\beta$. Then

$$e_\beta = K_\beta e_\beta + (K - K_\beta)x,$$

$$\|e_\beta\| \leqq \|(I - K_\beta)^{-1}\| \, \|(K - K_\beta)x\|,$$

which imply that $\{e_\beta\}$ is equicontinuous and bounded. By the Arzelà-Ascoli lemma, every sequence from $\{e_\beta\}$ has a subsequence which converges uniformly on each finite interval. Assume that $\beta_i \to \infty$ and $e_{\beta_i}(s) \to e(s)$ uniformly on finite intervals. In the identity

$$(I - K_\beta)e_{\beta_i} = (K - K_{\beta_i})x + (K_{\beta_i} - K_\beta)e_{\beta_i}$$

first let $\beta_i \to \infty$ and then $\beta \to \infty$ to get $(I - K)e = 0$. Then $e = 0$ since $(I - K)^{-1}$ exists. Thus, all limits of convergent sequences from $\{e_\beta\}$ are zero. It follows that $e_\beta(s) \to 0$ and, hence,

$$x_\beta(s) \to x(s) \quad \text{as} \quad \beta \to \infty,$$

uniformly for s in any finite interval. As noted above, the collectively compact operator approximation theory can be used to approximate x_β.

For numerical results and a discussion of computational procedures, see [22].

46

3.7 VARIANTS OF THE APPROXIMATION THEORY

Again let $\mathscr{X}$ be a Banach space and suppose that operators $K, K_n \in [\mathscr{X}]$ satisfy the basic hypotheses: $K_n \to K$, $\{K_n\}$ is collectively compact, and K is compact. Also assume that $(I - K)^{-1}$ exists. Then $(I - K_n)^{-1}$ exists for all n sufficiently large and

$$(I - K_n)^{-1} \to (I - K)^{-1}.$$

In the application of the approximation theory to the numerical solution of integral equations, given in Chapter 2, the nth approximation involves an $n \times n$ matrix. Thus, prohibitively large algebraic systems may be required in order to achieve a desired accuracy. Several ways to circumvent this difficulty will be presented. Without loss of generality, suppose that $(I - K_n)^{-1}$ exists for all n.

As approximations to

$$(I - K)^{-1} = I + (I - K)^{-1}K$$

consider (cf. Theorem 1.10 and also [47], pg. 552)

$$I + (I - K_n)^{-1}K, \qquad n = 1, 2, \ldots .$$

Note that

$$[I + (I - K_n)^{-1}K] - (I - K)^{-1} = (I - K_n)^{-1}(K_n - K)K(I - K)^{-1}.$$

By Theorem 1.6 and Corollary 1.9,

$$\| [I + (I - K_n)^{-1}K] - (I - K)^{-1} \| \to 0.$$

Hence,

$$[I + (I - K_n)^{-1}K]y \to (I - K)^{-1}y$$

uniformly for y in any bounded subset of $\mathscr{X}$. In contrast,

$$(I - K_n)^{-1}y \to (I - K)^{-1}y$$

uniformly for y in any compact set. In the applications to integral equations, the latter convergence rate depends on the magnitude and smoothness of y, whereas the former convergence rate depends only on the magnitude of y. Similar remarks pertain to other operator norm approximations of $(I - K)^{-1}$ derived below.

47

The operators $(I - K)^{-1}$ and $I + (I - K_n)^{-1}K$ are related by

$$(I - A_n)(I - K)^{-1} = I + (I - K_n)^{-1}K,$$

where

$$A_n = (I - K_n)^{-1}(K - K_n)K, \qquad \|A_n\| \to 0.$$

Fix n such that

$$\|A_n\| < 1, \qquad (I - A_n)^{-1} \in [\mathscr{X}],$$

$$(I - K)^{-1} = (I - A_n)^{-1}[I + (I - K_n)^{-1}K].$$

For $j = 0, 1, 2, \ldots$,

$$(I - A_n)^{-1} = I + A_n + \cdots + A_n^j + (I - A_n)^{-1}A_n^{j+1}.$$

Define approximations to $(I - K)^{-1}$ by

$$B_{nj} = (I + A_n + \cdots + A_n^j)[I + (I - K_n)^{-1}K]$$

or, equivalently, by

$$B_{n0} = I + (I - K_n)^{-1}K,$$

$$B_{nj} = B_{n0} + A_n B_{n,j-1}.$$

Then

$$\|B_{nj} - (I - K)^{-1}\| \leq \|(I - A_n)^{-1}\| \, \|A_n\|^{j+1} \, \|B_{n0}\|,$$

$$\|B_{nj} - (I - K)^{-1}\| \to 0 \quad \text{as} \quad j \to \infty.$$

In the integral equations applications, the determination of the successive approximations B_{nj}, $j = 0, 1, 2, \ldots$, involve multiplications by the fixed $n \times n$ matrix corresponding to $(I - K_n)^{-1}$. The operator K in B_{n0} usually would have to be replaced by some sufficiently accurate discrete approximation.

Except for differences in organization, the foregoing approximation method was proposed by Atkinson [23] along with variations on the theme and a discussion of practical implementation.

Earlier, Brakhage [25] used essentially the same formulas to obtain successive approximations to $(I - K_m)^{-1}$ from $(I - K_n)^{-1}$, where $m > n$ and n is fixed. All we need to do is to replace K by K_m in the preceding paragraph. In Brakhage's approach, the problem is completely discretized and thus amenable for calculation. As above, only a single matrix inverse, that corresponding to $(I - K_n)^{-1}$, is needed, no matter how large m is. It is a simple matter to combine the error bounds for the approximation of

$(I - K_m)^{-1}$ and for the discrepancy between $(I - K_m)^{-1}$ and $(I - K)^{-1}$ to obtain total error estimates.

Another method which leads to operator norm convergence was given by Anselone and Gonzalez-Fernandez [12]. It is based on the determination of operators $T,L \in [\mathcal{X}]$ such that $T^{-1} \in [\mathcal{X}]$, L is compact, and

$$T(I - K) = I - KL.$$

Then $(I - K)^{-1}$ exists iff $(I - KL)^{-1}$ exists, in which case

$$(I - K)^{-1} = (I - KL)^{-1}T.$$

We approximate $I - KL$ by $I - K_n L$. Then Proposition 1.8 yields

$$\|K_n L - KL\| \to 0.$$

Therefore, by Proposition 1.4, $(I - K)^{-1}$ exists iff for n sufficiently large there exist uniformly bounded $(I - K_n L)^{-1}$, in which case

$$\|(I - K_n L)^{-1}T - (I - K)^{-1}\| \to 0.$$

Error bounds are given in Proposition 1.3.

Such operators T and L can be found in a variety of ways. If $(I + K)^{-1}$ exists, we may take $T = I + K$ and $L = K$. More generally, for any prime number p consider

$$T_p = I + K + \cdots + K^{p-1},$$

$$T_p(I - K) = I - K^p = I - KL_p, \qquad L_p = K^{p-1}.$$

If p is sufficiently large then T_p^{-1} exists. We indicate the proof when $\mathcal{X}$ is complex; the real case follows by means of a complexification argument. Note that

$$T_p = \prod_{q=1}^{p-1} (K - \lambda_{pq}I),$$

where the λ_{pq} are the nontrivial p^{th} roots of unity. Hence, for $1 \leq q < p$ and p prime, the λ_{pq} are distinct numbers with absolute value one. Since K is compact, its eigenvalues form a finite set or a sequence converging to zero. Thus, only finitely many of the λ_{pq} can be eigenvalues of K. It follows that $T_p^{-1} \in [\mathcal{X}]$ for every sufficiently large prime p. (Cf. Appendix 1, Theorem 11.)

Other possibilities for T and L are

$$T = I + K + cK^2,$$

$$L = (1 - c)K + cK^2.$$

If $K^2 \neq O$ then there exists c such that $T^{-1} \in [\mathscr{X}]$. The reasoning is similar to that used in the preceding paragraph.

To illustrate the method under discussion, let K and L be integral operators on $\mathscr{C} = \mathscr{C}[0,1]$ with continuous kernels k and l. As before, define K_n, $n \geq 1$, by means of a convergent quadrature formula. Then (cf. Chapter 2, §2.5) KL and $K_n L$ are the integral operators with the continuous kernels

$$\varphi(k_s l^t) = \int_0^1 k(s,\tau) l(\tau,t)\, d\tau,$$

$$\varphi_n(k_s l^t) = \sum_{j=1}^n w_{nj} k(s,t_{nj}) l(t_{nj},t).$$

Hence,

$$\|K_n L - KL\| \leq \max_{0 \leq s,t \leq 1} |\varphi_n(k_s l^t) - \varphi(k_s l^t)|.$$

This inequality and Proposition 1.3 yield estimates of

$$\|(I - K_n L)^{-1} T - (I - K)^{-1}\|.$$

To determine $(I - K_n L)^{-1}$, it suffices to solve

$$x_n - K_n L x_n = y.$$

Operate with L and evaluate to obtain

$$(Lx_n)(t_{ni}) - (LK_n L x_n)(t_{ni}) = (Ly)(t_{ni}).$$

Now

$$(LK_n x)(t_{ni}) = \sum_{j=1}^n A_{ij}^n x(t_{nj}),$$

where

$$A_{ij}^n = w_{nj} \int_0^1 l(t_{ni},\tau) k(\tau,t_{nj})\, d\tau.$$

Therefore,

$$(Lx_n)(t_{ni}) - \sum_{j=1}^n A_{ij}^n (Lx_n)(t_{nj}) = (Ly)(t_{ni}).$$

If this system is solved for $(Lx_n)(t_{nj})$, $j = 1, \ldots, n$, then $K_n L x_n$ and $x_n = K_n L x_n + y$ can be evaluated. An explicit formula for $(I - K_n L)^{-1}$ can be written (cf. Chapter 2, §2.4).

The present method is not easily compared with that of Chapter 1. We now have norm convergence,

$$\|(I - K_n L)^{-1} T - (I - K)^{-1}\| \to 0,$$

50

instead of pointwise convergence,

$$(I - K_n)^{-1} \to (I - K)^{-1}.$$

However, the determination of $(I - K_n L)^{-1}$ is equivalent to a matrix problem in which each matrix element is an integral, whereas the elements of the matrix corresponding to $I - K_n$ are function values (cf. §3.10). Another difference between the two methods lies in the ad hoc way T and L are found in the latter.

Physical problems sometimes lead rather directly to integral equations of the form $(I - KL)x = y$. In such cases we can proceed immediately to the approximations $I - K_n L$. Examples are furnished by the bending of a beam [71], the diffraction of electromagnetic pulses by a dielectric wedge [66], and by the scattering of radiation and neutrons, which is discussed in the next section.

3.8 APPROXIMATE SOLUTIONS OF TRANSPORT EQUATIONS

The distribution of radiation or neutrons in a scattering-emitting-absorbing medium is governed by an integrodifferential equation known variously as the transport, transfer, and Boltzmann equation [31]. In the discrete-ordinates method, numerical integration yields approximate solutions. The convergence of these approximations was established in a number of cases by Anselone [1–5]. Although that work preceded by several years the formulation of the collectively compact operator approximation theory, it does anticipate the later theory in many respects.

We shall discuss the convergence of the discrete-ordinates method for the relatively simple transport problem:

$$\mu \frac{\partial f(\tau,\mu)}{\partial \tau} = f(\tau,\mu) - \tfrac{1}{2} \int_{-1}^{1} f(\tau,\mu')\, d\mu' - e^{-\tau}, \qquad \begin{array}{l} 0 \leq \tau \leq 1, \\ -1 \leq \mu \leq 1, \end{array}$$

$$f(0,\mu) = 0 \quad \text{for} \quad \mu < 0, \qquad f(1,\mu) = 0 \quad \text{for} \quad \mu > 0.$$

Here $f(\tau,\mu)$ is intensity at (optical) distance τ from the surface $\tau = 0$ of a slab and in a direction making an angle $\theta = \cos^{-1} \mu$ with the outward normal at $\tau = 0$. There is no absorption. Scattering is isotropic. The source term $e^{-\tau}$ comes from the scattering of a parallel beam incident normally at $\tau = 0$.

The problem for $f(\tau,\mu)$ is transformed into a Fredholm integral equation as follows. Let

$$x(\tau) = \tfrac{1}{2} \int_{-1}^{1} f(\tau,\mu)\, d\mu + e^{-\tau}.$$

Then

$$\mu\, \frac{\partial f(\tau,\mu)}{\partial \tau} = f(\tau,\mu) - x(\tau).$$

Solve for $f(\tau,\mu)$ in terms of $x(\tau)$:

$$f(\tau,\mu) = \int_{0}^{\tau} e^{(\tau-t)/\mu} x(t)\, \frac{dt}{-\mu}, \qquad \mu < 0,$$

$$f(\tau,0) = x(\tau),$$

$$f(\tau,\mu) = \int_{\tau}^{1} e^{(\tau-t)/\mu} x(t)\, \frac{dt}{\mu}, \qquad \mu > 0.$$

Substitute into the defining equation for $x(\tau)$ and simplify to obtain

$$x(\tau) - \tfrac{1}{2} \int_{0}^{1} E_1(|\tau - t|) x(t)\, dt = e^{-\tau},$$

where E_1 is the exponential integral function,

$$E_1(s) = \int_{0}^{1} e^{-s/\mu} \mu^{-1} d\mu, \qquad s > 0,$$

which has a logarithmic singularity at $s = 0$.

Let K be the integral operator on $\mathscr{C} = \mathscr{C}[0,1]$ with the kernel $\tfrac{1}{2} E_1(|\tau - t|)$. Then

$$(I - K)x = y, \qquad y(\tau) = e^{-\tau}.$$

It is not difficult to show that $\|K\| < 1$. Therefore, $(I - K)^{-1} \in [\mathscr{C}]$ and x is given by

$$x = \sum_{m=0}^{\infty} K^m y.$$

Substitution into the equations for $f(\tau,\mu)$ in terms of $x(\tau)$ yields the solution of the problem posed.

Although the foregoing approach does establish the existence and uniqueness of $f(\tau,\mu)$, it has limited practical value. A more efficient numerical procedure is the discrete-ordinates method based on the equations

$$\mu \frac{df_n(\tau,\mu)}{d\tau} = f_n(\tau,\mu) - \tfrac{1}{2} \sum_{j=\pm 1}^{\pm n} w_{nj} f_n(\tau,\mu_{nj}) - e^{-\tau}, \qquad \begin{array}{c} 0 \leq \tau \leq 1, \\ -1 \leq \mu \leq 1, \end{array}$$

$$f_n(0,\mu) = 0 \quad \text{for} \quad \mu < 0, \qquad f_n(1,\mu) = 0 \quad \text{for} \quad \mu > 0,$$

where $n = 1, 2, \ldots$ and, for example, the Gauss quadrature formula is employed. For each n, the functions $f_n(\tau,\mu_{ni})$, $i = \pm 1, \ldots, \pm n$ satisfy a system of ordinary differential equations which is readily solved. Then the function

$$x_n(\tau) = \tfrac{1}{2} \sum_{j=\pm 1}^{\pm n} w_{nj} f_n(\tau,\mu_{nj}) + e^{-\tau}$$

is determined. Since

$$\mu \frac{df_n(\tau,\mu)}{d\tau} = f_n(\tau,\mu) - x_n(\tau),$$

$f_n(\tau,\mu)$ is available from

$$f_n(\tau,\mu) = \int_0^\tau e^{(\tau-t)/\mu} x_n(t) \frac{dt}{-\mu}, \qquad \mu < 0,$$

$$f_n(\tau,0) = x_n(\tau),$$

$$f_n(\tau,\mu) = \int_\tau^1 e^{(\tau-t)/\mu} x_n(t) \frac{dt}{\mu}, \qquad \mu > 0.$$

See Chandrasekhar [31] for explicit formulas and numerical calculations.

The convergence of the discrete-ordinates approximations was established as follows. The substitution of the foregoing equations for $f_n(\tau,\mu)$ into the defining equation for $x_n(\tau)$ gives

$$x_n(\tau) - \tfrac{1}{2} \int_0^1 E_{n1}(|\tau - t|) x_n(t) \, dt = e^{-\tau},$$

where E_{n1} is a quadrature approximation of E_1:

$$E_{n1}(s) = \sum_{j=1}^n w_{nj} e^{-s/\mu_{nj}} \mu_{nj}^{-1}, \qquad s > 0.$$

From the properties of the Gauss quadrature formula,

$$E_{n1}(s) \to E_1(s) \quad \text{as} \quad n \to \infty, \qquad s > 0,$$

uniformly on $[\varepsilon,1]$ for each $\varepsilon > 0$. Let K_n be the integral operator on $\mathscr{C}$

with the kernel $\frac{1}{2}E_{n1}(|\tau - t|)$. Then

$$(I - K_n)x_n = y, \qquad y(\tau) = e^{-\tau},$$
$$\|K_n - K\| \to 0,$$
$$\|(I - K_n)^{-1} - (I - K)^{-1}\| \to 0,$$
$$\|x_n - x\| \to 0,$$
$$f_n(\tau,\mu) \to f(\tau,\mu) \quad \text{uniformly.}$$

The singularity of E_1 slightly complicates the demonstrations of the conclusions.

The equations relating $x(\tau)$ with $f(\tau,\mu)$ and $x_n(\tau)$ with $f_n(\tau,\mu)$ can be expressed in operator form:

$$x = Mf + y, \qquad f = Lx,$$
$$x_n = M_n f_n + y, \qquad f_n = Lx_n.$$

A glance at the derivations of K and K_n reveals that

$$K = ML, \qquad K_n = M_n L.$$

Although it was not done that way originally, the convergence $\|K_n - K\| \to 0$ could be established by arguments used in §3.7.

A much more difficult transport problem pertains to a semi-infinite slab $(0 \leq t < \infty)$ in which there is isotropic scattering, no emission, no absorption, and no source term. These and other physical assumptions lead to a homogeneous integral equation

$$x(\tau) - \tfrac{1}{2} \int_0^\infty E_1(|\tau - t|)x(t)\, dt = 0.$$

The Wiener-Hopf method was designed to solve equations of this general type by means of Fourier transforms. A more direct method, based on monotonicity arguments, was given by Hopf [44]. It involves a change of variable

$$x(\tau) = \tau + q(\tau).$$

Then q satisfies the inhomogeneous equation

$$q(\tau) - \tfrac{1}{2} \int_0^\infty E_1(|\tau - t|)q(t)\, dt = E_3(\tau),$$

where

$$E_3(\tau) = \int_0^1 e^{-\tau/\mu}\mu\, d\mu.$$

Let K denote the integral operator with kernel $\frac{1}{2}E_1(|\tau - t|)$ defined on the space of bounded continuous functions on $[0,\infty)$ equipped with the supremum norm. Then

$$(I - K)q = E_3.$$

Although $\|K\| = 1$, positivity and monotonicity considerations yield a unique positive solution q, which is given by the Neumann series

$$q = \sum_{m=0}^{\infty} K^m E_3.$$

The problem for the discrete-ordinates approximations can be recast in the form

$$x_n(\tau) - \frac{1}{2}\int_0^{\infty} E_{n1}(|\tau - t|)x_n(t)\, dt = 0.$$

Let

$$x_n(\tau) = \tau + q_n(\tau).$$

Then

$$(I - K_n)q_n = E_{n3},$$

where K_n is the integral operator with kernel $\frac{1}{2}E_{n1}(|\tau - t|)$ and E_{n3} is a quadrature approximation of E_3. The function q_n is given by

$$q_n = \sum_{m=0}^{\infty} K_n^m E_{n3}.$$

A detailed term by term analysis [1–3] of the series for q and q_n yields

$$\|q_n - q\| \to 0,$$

from which the convergence of the discrete-ordinates approximations follows.

Some of the results obtained for isotropic transport problems were extended by Nestell [62] to cases with anisotropic scattering.

3.9   BOUNDARY VALUE PROBLEMS FOR PARTIAL DIFFERENTIAL EQUATIONS

Some very interesting applications of the collectively compact approximation theory to boundary value problems, recast in singular integral equation form, have appeared too late for a detailed synopsis to be prepared. We merely cite the references: Gilbert [77], Gilbert and Colton [78], Atkinson [79].

3.10   KERNELS OF FINITE RANK

Consider an integral operator $K \in [\mathscr{C}]$ of the form

$$(Kx)(s) = \int_0^1 k(s,t)x(t)\, dt, \qquad k(s,t) = \sum_{\alpha=1}^{m} g^\alpha(s)h^\alpha(t),$$

with $\{g^\alpha\}$ and $\{h^\alpha\}$ linearly independent. Then [48, 71] $x - Kx = y$ iff

$$x = y + \sum_{\alpha=1}^{m} x_\alpha g^\alpha, \qquad x_\alpha - \sum_{\beta=1}^{m} c_{\alpha\beta} x_\beta = y_\alpha,$$

$$y_\alpha = \int_0^1 h^\alpha(t)y(t)\, dt, \qquad c_{\alpha\beta} = \int_0^1 h^\alpha(t)g^\beta(t)\, dt.$$

This reduces the integral equation $x - Kx = y$ to a matrix problem. Usually, y_α and $c_{\alpha\beta}$ must be approximated by numerical integration. It turns out that, for K_n defined as in §2.1, $x - K_n x = y$ iff

$$x = y + \sum_{\alpha=1}^{m} x_\alpha g^\alpha, \qquad x_\alpha - \sum_{\beta=1}^{m} c_{\alpha\beta} x_\beta = y_\alpha,$$

$$y_\alpha = \sum_{j=1}^{n} w_{nj} h^\alpha(t_{nj})y(t_{nj}), \qquad c_{\alpha\beta} = \sum_{j=1}^{n} w_{nj} h^\alpha(t_{nj})g^\beta(t_{nj}).$$

So the collectively compact theory gives convergence criteria and error bounds. Curiously, $x - K_n x = y$ reduces to an $m \times m$ matrix problem as above, and to an $n \times n$ matrix problem as in §2.1.

Now let $K \in [\mathscr{C}]$ be any integral operator with a continuous kernel k. A popular scheme is to construct integral operators K^m with kernels of finite rank k^m such that $k^m \to k$ uniformly and, hence $\|K^m - K\| \to 0$. Proposition 1.4 relates $x - Kx = y$ and $x - K^m x = y$, which can be treated as in the preceding paragraph.

chapter **4**

SPECTRAL APPROXIMATIONS

4.1 INTRODUCTION AND SUMMARY

Again let $\mathscr{X}$ be a real or complex Banach space, $\mathscr{B}$ the closed unit ball in $\mathscr{X}$, and $[\mathscr{X}]$ the space of bounded linear operators on $\mathscr{X}$. We shall compare spectral properties of operators $T, T_n \in [\mathscr{X}]$, $n = 1, 2, \ldots$, such that

$$T_n \to T, \qquad \{T_n - T\} \text{ is collectively compact.}$$

In view of the final paragraph of Chapter 1, $\{T_n - T\}$ is collectively compact iff $\{(T_n - T)x_n\}$ is relatively compact for each sequence $\{x_n\} \subset \mathscr{B}$.

The following proposition indicates that the present hypotheses are implied by those assumed in Chapter 1.

PROPOSITION 4.1. Let $T, T_n \in [\mathscr{X}]$, $n = 1, 2, \ldots$. Then the following four sets of conditions are equivalent:

$\{T_n\}$ collectively compact, T compact;

$\{T_n - T\}$ collectively compact, T compact;

$\{T_n - T\}$ collectively compact, some T_n compact;

$\{T_n - T\}$ collectively compact, every T_n compact.

57

Since the proof is easy it is omitted.

The following standard terminology and notation will be employed. For $T \in [\mathscr{X}]$ let $\mathscr{N}(T) = \{x \in \mathscr{X} \colon Tx = 0\}$, the *null space* of T. Scalars (real or complex) will be denoted by λ and μ, sometimes with subscripts. For brevity, $\lambda I - T = \lambda - T$. A scalar λ is an *eigenvalue* of T iff $\mathscr{N}(\lambda - T) \neq \Theta$, in which case $\mathscr{N}(\lambda - T)$ is the corresponding *eigenmanifold* and $\mathscr{N}[(\lambda - T)^k]$, $k = 1, 2, \ldots$, are *generalized eigenmanifolds*. The *resolvent set* for T is

$$\rho(T) = \{\lambda \colon \exists (\lambda - T)^{-1} \in [\mathscr{X}]\},$$

the *spectrum* $\sigma(T)$ is the complement of $\rho(T)$, and the *point spectrum* $\sigma_p(T) \subset \sigma(T)$ consists of the eigenvalues of T. General properties of the resolvent set and spectrum are reviewed in §4.3. Other concepts, such as spectral sets, spectral projections, spectral subspaces, and functions of operators are introduced later in the chapter.

For operators $T, T_n \in [\mathscr{X}]$ such that $T_n \to T$ and $\{T_n - T\}$ is collectively compact, it will be shown that any prescribed neighborhood of $\sigma(T)$ contains $\sigma(T_n)$ for all n sufficiently large. If $\mathscr{X}$ is complex, and an operator $f(T)$ is defined by the operational calculus (cf. §4.7), then $f(T_n)$ is defined for all n sufficiently large, $f(T_n) \to f(T)$, and $\{f(T_n) - f(T)\}$ is collectively compact. In the important case with $f(T)$ and $f(T_n)$ spectral projections, the corresponding spectral subspaces eventually have the same dimension. Other results compare eigenvalues, eigenvectors, and generalized eigenvectors of T and T_n.

References for this chapter are [15, 16] by Anslone and Palmer, and [21] by Atkinson. The paper by Atkinson concerns the case where the operators satisfy the hypotheses in Chapter 1. For related work see Brakhage [26], Bückner [28–30] and Wielandt [74].

4.2 PROPERTIES OF COLLECTIVELY COMPACT SETS

Various properties of collectively compact sets will be needed in the subsequent analysis. For example, any subset or scalar multiple of a collectively compact set is collectively compact. Any finite union or finite sum of collectively compact sets is collectively compact. Further properties are given in the following proposition.

PROPOSITION 4.2. Let $\mathscr{K} \subset [\mathscr{X}]$ be collectively compact. Then each of the following sets is collectively compact:

1. $\Lambda\mathscr{K} = \{\lambda K: \lambda \in \Lambda,\ K \in \mathscr{K}\}$ for any bounded scalar set Λ.
2. $\mathscr{K}\mathscr{M} = \{KM: K \in \mathscr{K},\ M \in \mathscr{M}\}$ for any bounded set $\mathscr{M} \subset [\mathscr{X}]$.
3. $\mathscr{N}\mathscr{K} = \{NK: N \in \mathscr{N},\ K \in \mathscr{K}\}$ for any relatively compact set $\mathscr{N} \subset [\mathscr{X}]$.
4. The norm closure $\overline{\mathscr{K}}$ and strong closure $\overline{\mathscr{K}}^s$ of $\mathscr{K}$.
5. c.h. $\mathscr{K}$, the convex hull of $\mathscr{K}$.
6. $\left\{\sum\limits_{j=1}^{J} \lambda_j K_j: K_j \in \mathscr{K}, \sum\limits_{j=1}^{J} |\lambda_j| \leq b\right\}$ for any $b < \infty$ and any $J \leq \infty$.
7. $\{\int_\Gamma K(\lambda)\,d\lambda: K(\lambda) \in \mathscr{K},\ \ell(\Gamma) \leq b\}$ for any $b < \infty$, where Γ is an interval or rectifiable arc of length $\ell(\Gamma)$ and the integrals are strong or norm limits of the usual approximating sums

$$\sum_{j=1}^{J} K(\lambda_j')(\lambda_j - \lambda_{j-1}).$$

Proof. Only the main steps in the arguments are given.

1. $\Lambda\mathscr{K}\mathscr{B} \subset r\mathscr{K}\mathscr{B}$ for $r = \sup\{|\lambda|: \lambda \in \Lambda\}$.
2. $\mathscr{K}\mathscr{M}\mathscr{B} \subset r\mathscr{K}\mathscr{B}$ for $r = \sup\{\|T\|: T \in \mathscr{M}\}$.
3. Define $f: [\mathscr{X}] \times \mathscr{X} \to \mathscr{X}$ by $f(T,y) = Ty$. Then f is uniformly continuous on bounded sets, $\mathscr{N} \times \mathscr{K}\mathscr{B}$ is relatively compact, and $\mathscr{N}\mathscr{K}\mathscr{B} = f(\mathscr{N} \times \mathscr{K}\mathscr{B})$ is relatively compact.
4. $\overline{\mathscr{K}}\mathscr{B} \subset \overline{\mathscr{K}}^s\mathscr{B} \subset \overline{\mathscr{K}\mathscr{B}}$.
5. (c.h.$\mathscr{K}$)$\mathscr{B} \subset$ c.h.($\mathscr{K}\mathscr{B}$), which is relatively compact by Mazur's theorem [38, p. 416].
6. This follows from 1, 5, and 4 (needed when $J = \infty$).
7. This is a consequence of 4, 6, and

$$\sum_{j=1}^{J} |\lambda_j - \lambda_{j-1}| \leq \ell(\Gamma).$$

There are generalizations of 1–4 for operators from one normed linear space (not necessarily complete) to another [15]. A recent counterexample due to J. A. Higgins (private communication) implies that the completeness of the range space is essential in 5 and, hence, also in 6 and 7.

4.3 RESOLVENT SETS AND SPECTRA

Let us review some of the basic properties of the resolvent set $\rho(T)$ and the spectrum $\sigma(T)$ of an operator $T \in [\mathscr{X}]$. Standard references are Dunford and Schwartz [38], Taylor [70], and Yosida [75].

By Proposition 1.3, if $\lambda \in \rho(T)$ and $|\mu - \lambda| < \|(\lambda - T)^{-1}\|^{-1}$, then $\mu \in \rho(T)$. Therefore, $\rho(T)$ is open, $\sigma(T)$ is closed, and

$$\|(\lambda - T)^{-1}\| \text{ dist. } (\lambda, \sigma(T)) \geq 1 \quad \text{for} \quad \lambda \in \rho(T).$$

By Proposition 1.1, if $|\lambda| > \|T\|$, then $\lambda \in \rho(T)$ and

$$(\lambda - T)^{-1} = \sum_{k=0}^{\infty} \frac{T^k}{\lambda^{k+1}},$$

$$\|(\lambda - T)^{-1}\| \leq \frac{1}{|\lambda| - \|T\|},$$

$$\|(\lambda - T)^{-1}\| \to 0 \quad \text{as} \quad |\lambda| \to \infty.$$

Hence, $|\lambda| \leq \|T\|$ for $\lambda \in \sigma(T)$, $\sigma(T)$ is bounded, and $\sigma(T)$ is compact.

A more detailed analysis yields the following sharper results. For each $T \in [\mathscr{X}]$, there exists $r(T) = \lim\limits_{k \to \infty} \|T^k\|^{1/k}$, $r(T) \leq \|T\|$, $|\lambda| \leq r(T)$ for $\lambda \in \sigma(T)$, and $r(T)$ is the radius of convergence of the power series for $(\lambda - T)^{-1}$ displayed above, i.e., the series converges for $|\lambda| > r(T)$ and diverges for $|\lambda| < r(T)$. If $\mathscr{X}$ is complex and $\mathscr{X} \neq \Theta$, then $\sigma(T)$ is nonvoid and

$$r(T) = \max \{|\lambda|: \lambda \in \sigma(T)\}.$$

It will be convenient to compactify the scalar field with the point $\lambda = \infty$ when $\mathscr{X}$ is complex and with $\lambda = \pm\infty$ when $\mathscr{X}$ is real. Define $(\lambda - T)^{-1} = O$ for such λ and any $T \in [\mathscr{X}]$. The *extended resolvent set* $\tilde{\rho}(T)$ consists of $\rho(T)$ and the compactification point(s). In the extended scalar field a set is compact iff it is closed.

The *resolvent* for $T \in [\mathscr{X}]$ is the function $\lambda \mapsto (\lambda - T)^{-1}$ from $\tilde{\rho}(T)$ to $[\mathscr{X}]$. By Proposition 1.3, it is a continuous function (in fact, it is analytic). Therefore, $\{(\lambda - T)^{-1}: \lambda \in \Lambda\}$ is compact for each closed (= compact) set

$\Lambda \subset \tilde{\rho}(T)$. The *resolvent equation,*

$$(\lambda - T)^{-1} - (\mu - T)^{-1} = (\mu - \lambda)(\lambda - T)^{-1}(\mu - T)^{-1}, \qquad \lambda, \mu \in \rho(T),$$

is frequently useful.

4.4   SPECTRAL APPROXIMATION THEOREMS

The results in this section relate the spectra and the resolvents of operators T and T_n which satisfy the hypotheses in §4.1. For the first few theorems, λ is a fixed scalar. The initial result generalizes Theorem 1.6, which will be used in the proof.

THEOREM 4.3. Let $T, T_n \in [\mathscr{X}]$, $n = 1, 2, \ldots$. Assume $T_n \to T$ and $\{T_n - T\}$ collectively compact. Then

     (a) $\lambda \in \rho(T)$

iff (b) there exists N such that $\lambda \in \rho(T_n)$ for $n \geq N$ and the set $\{(\lambda - T_n)^{-1} : n \geq N\}$ is bounded,

in which case

     (c) $(\lambda - T_n)^{-1} \to (\lambda - T)^{-1}$,

     (d) $\{(\lambda - T_n)^{-1} - (\lambda - T)^{-1} : n \geq N\}$ is collectively compact.

Proof. First assume $\lambda \in \rho(T)$. Then

$$\lambda - T_n = (I - K_n)(\lambda - T), \qquad K_n = (T_n - T)(\lambda - T)^{-1},$$

$$K_n \to O, \qquad \{K_n\} \text{ is collectively compact.}$$

By Theorem 1.6 with $K = O$, there exists N such that $(I - K_n)^{-1} \in [\mathscr{X}]$ for $n \geq N$, $\{(I - K_n)^{-1} : n \geq N\}$ is bounded, and $(I - K_n)^{-1} \to I$. Consequently, $\lambda \in \rho(T_n)$ for $n \geq N$ and

$$(\lambda - T_n)^{-1} = (\lambda - T)^{-1}(I - K_n)^{-1} \to (\lambda - T)^{-1},$$

$$(\lambda - T_n)^{-1} - (\lambda - T)^{-1} = (\lambda - T)^{-1}(T_n - T)(\lambda - T_n)^{-1}.$$

Thus, by Proposition 4.2, (a) $\Rightarrow$ (b), (c), (d).

Now assume (b). By Lemma 1.5, $(\lambda - T)^{-1}$ exists as an operator defined on $(\lambda - T)\mathcal{X}$. For $n \geq N$,

$$(\lambda - T) = (I - L_n)(\lambda - T_n), \qquad L_n = (T - T_n)(\lambda - T_n)^{-1},$$
$$(I - L_n)^{-1} \text{ exists}, \qquad\qquad L_n \text{ is compact.}$$

By the Fredholm alternative, $(I - L_n)^{-1} \in [\mathcal{X}]$ and

$$(\lambda - T)^{-1} = (\lambda - T_n)^{-1}(I - L_n)^{-1} \in [\mathcal{X}], \qquad n \geq N.$$

Thus, (b) $\Rightarrow$ (a) and the proof is complete.

In (b), suppose that

$$\|(\lambda - T_n)^{-1}\| \leq r < \infty, \qquad n \geq N.$$

Then

$$\text{dist. } (\lambda, \sigma(T_n)) \geq \frac{1}{r}, \qquad n \geq N,$$

$$\mu \in \rho(T_n) \quad \text{for} \quad |\mu - \lambda| < \frac{1}{r}, \qquad n \geq N.$$

Thus, (b) implies that some neighborhood of λ belongs to $\rho(T_n)$ for every $n \geq N$.

Somewhat longer proofs of the separate parts of Theorem 4.3, which also yield useful bounds, will be given. The analysis is based on the following well known consequence of Proposition 1.1.

PROPOSITION 4.4. Let $A \in [\mathcal{X}]$ and $\|A^2\| < 1$. Then there exists $(I - A)^{-1} \in [\mathcal{X}]$ and

$$\|(I - A)^{-1}\| \leq \frac{\|I + A\|}{1 - \|A^2\|}.$$

The next theorem yields another proof that (a) $\Rightarrow$ (b), (c) in Theorem 4.3.

THEOREM 4.5. $T, T_n \in [\mathcal{X}]$, $n = 1, 2, \ldots$. Assume $T_n \to T$ and $\{T_n - T\}$ collectively compact. Let $\lambda \in \rho(T)$ and

$$K_n = (T_n - T)(\lambda - T)^{-1}.$$

Then $K_n \to O$ and $\|K_n^2\| \to 0$. Whenever $\|K_n^2\| < 1$, then $\lambda \in \rho(T_n)$,

$$\|(\lambda - T_n)^{-1}\| \leq \frac{\|(\lambda - T)^{-1}\| \, \|I + K_n\|}{1 - \|K_n^2\|},$$

and

$$\|(\lambda - T_n)^{-1}x - (\lambda - T)^{-1}x\| \leq \|(\lambda - T_n)^{-1}\| \, \|K_n x\| \to 0, \qquad x \in \mathcal{X}.$$

Proof. As we observed earlier,

$$\lambda - T_n = (I - K_n)(\lambda - T),$$

$$K_n \to O, \qquad \{K_n\} \text{ is collectively compact.}$$

By Corollary 1.9, $\|K_n^2\| \to 0$. Choose any n such that $\|K_n^2\| < 1$. By Proposition 4.4,

$$(I - K_n)^{-1} \in [\mathcal{X}],$$

$$(\lambda - T_n)^{-1} = (\lambda - T)^{-1}(I - K_n)^{-1} \in [\mathcal{X}],$$

$$(\lambda - T_n)^{-1} - (\lambda - T)^{-1} = (\lambda - T_n)^{-1}K_n,$$

and the desired results follow.

The next theorem yields (b) $\Rightarrow$ (a) in Theorem 4.3, plus a criterion for $\lambda \in \rho(T)$ based on T_n for a single n.

THEOREM 4.6. Let $T, T_n \in [\mathcal{X}]$, $n = 1, 2, \ldots$. Assume $T_n \to T$ and $\{T_n - T\}$ collectively compact. With λ and n fixed, let $\lambda \in \rho(T_n)$ and

$$L_n = (T - T_n)(\lambda - T_n)^{-1}.$$

If $\|L_n^2\| < 1$, then $\lambda \in \rho(T)$,

$$\|(\lambda - T)^{-1}\| \leq \frac{\|(\lambda - T_n)^{-1}\| \, \|I + L_n\|}{1 - \|L_n^2\|},$$

and

$$\|(\lambda - T_n)^{-1}x - (\lambda - T)^{-1}x\| \leq \|(\lambda - T)^{-1}\| \, \|L_n x\| \to 0, \qquad x \in \mathcal{X}.$$

Now assume (b) of Theorem 4.3. Then $L_n \to O$, $\|L_n^2\| \to 0$, and the foregoing inequalities hold for all n sufficiently large.

Proof. The argument parallels that for Theorem 4.5, except for $L_n \to O$, which is a consequence of $K_n \to O$ and the identity

$$L_n = (L_n - I)K_n.$$

Theorems 4.5 and 4.6 yield the promised second demonstration of Theorem 4.3.

Next we allow λ to vary in a closed subset of $\tilde{\rho}(T)$.

THEOREM 4.7. Let $T, T_n \in [\mathscr{X}]$, $n = 1, 2, \ldots$. Assume $T_n \to T$ and $\{T_n - T\}$ collectively compact. Then for each closed set $\Lambda \subset \tilde{\rho}(T)$ there exists N such that:

$\Lambda \subset \tilde{\rho}(T_n)$ for $n \geq N$;
$\{(\lambda - T_n)^{-1} \colon \lambda \in \Lambda, n \geq N\}$ is bounded;
for each $x \in \mathscr{X}$, $(\lambda - T_n)^{-1}x \to (\lambda - T)^{-1}x$ uniformly for $\lambda \in \Lambda$;
$\{(\lambda - T_n)^{-1} - (\lambda - T)^{-1} \colon \lambda \in \Lambda, n \geq N\}$ is collectively compact;
the resolvents $\lambda \mapsto (\lambda - T_n)^{-1}, n \geq N$, are equicontinuous on Λ.

Proof. Let $K_n(\lambda) = (T_n - T)(\lambda - T)^{-1}$ for $n \geq 1$ and $\lambda \in \Lambda$. For each $\lambda \in \Lambda$, $K_n(\lambda) \to O$ and $\|K_n(\lambda)^2\| \to 0$, as above. For each $x \in \mathscr{X}$, the functions

$$\lambda \mapsto K_n(\lambda), \qquad \lambda \mapsto K_n(\lambda)x, \qquad \lambda \mapsto K_n(\lambda)^2, \qquad n \geq 1,$$

are equicontinuous on Λ. Therefore, since Λ is compact,
$\{K_n(\lambda) \colon \lambda \in \Lambda, n \geq 1\}$ is bounded,
for each $x \in \mathscr{X}$, $K_n(\lambda)x \to 0$ uniformly for $\lambda \in \Lambda$,
$\|K_n(\lambda)^2\| \to 0$ uniformly for $\lambda \in \Lambda$.
Now Theorem 4.5 gives the first three assertions. The fourth is implied by Proposition 4.2 and

$$(\lambda - T_n)^{-1} - (\lambda - T)^{-1} = (\lambda - T)^{-1}(T_n - T)(\lambda - T_n)^{-1}.$$

The final assertion follows from

$$(\lambda - T_n)^{-1} - (\mu - T_n)^{-1} = (\mu - \lambda)(\lambda - T_n)^{-1}(\mu - T_n)^{-1},$$

$$\lambda, \mu \in \rho(T_n),$$

and (to deal with $|\lambda| = \infty$) the bound for $\|(\lambda - T_n)^{-1}\|$ in Theorem 4.5.

Theorem 4.7 has a converse (analogous to Theorem 4.6) which we omit. The following result is an immediate consequence of Theorem 4.7.

THEOREM 4.8. Let $T, T_n \in [\mathscr{X}]$, $n = 1, 2, \ldots$. Assume $T_n \to T$ and $\{T_n - T\}$ collectively compact. Then for each open set $\Omega \supset \sigma(T)$ there exists N such that $\Omega \supset \sigma(T_n)$ for $n \geq N$.

4.5 EIGENVALUES AND EIGENVECTORS OF COMPACT OPERATORS

To illustrate Theorem 4.8 consider operators $T, T_n \in [\mathscr{X}]$, $n = 1, 2, \ldots$, such that $T_n \to T$, $\{T_n\}$ is collectively compact and, hence, T is compact. Then $\sigma(T)$ is finite or countable and 0 is the only possible limit point of $\sigma(T)$. If $\mathscr{X}$ is infinite dimensional, then $0 \in \sigma(T)$. Each nonzero $\mu \in \sigma(T)$ is an eigenvalue of T and the corresponding eigenmanifold $\mathscr{N}(\mu - T)$ is finite dimensional. The operators T_n have the same properties since they are compact. (Cf. Appendix 1.)

For each $\varepsilon > 0$ let (see Figure 1)

$$\Omega_\varepsilon(T) = \bigcup_{\mu \in \sigma(T)} \Omega_\varepsilon(\mu), \qquad \Omega_\varepsilon(\mu) = \{\lambda \colon |\lambda - \mu| < \varepsilon\}.$$

By Theorem 4.8 there exists N_ε such that

$$\sigma(T_n) \subset \Omega_\varepsilon(T) \quad \text{for} \quad n \geq N_\varepsilon.$$

Suppose that $\sigma(T)$ contains at least two points. Fix $\mu \in \sigma(T)$, $\mu \neq 0$, and define

$$\varepsilon(\mu, T) = \tfrac{1}{2} \operatorname{dist.} (\mu, \sigma(T) - \{\mu\}).$$

Then

$$\Omega_\varepsilon(\mu) \cap \Omega_\varepsilon(\lambda) = \phi \quad \text{for} \quad \lambda \in \sigma(T) - \{\mu\}, \qquad \varepsilon \leq \varepsilon(\mu, T),$$
$$\Omega_\varepsilon(\mu) \cap \sigma(T) = \{\mu\} \quad \text{for} \quad \varepsilon \leq \varepsilon(\mu, T).$$

In particular, if $\dim \mathscr{X} = \infty$, then $0 \in \sigma(T)$ and

$$|\lambda| > \tfrac{1}{2}|\mu| \quad \text{for} \quad \lambda \in \Omega_\varepsilon(\mu), \qquad \varepsilon \leq \varepsilon(\mu, T).$$

Let k_n be the (necessarily finite) number of points in $\sigma(T_n) \cap \Omega_\varepsilon(\mu)$, where $\varepsilon = \varepsilon(\mu, T)$. Whenever $k_n \neq 0$, we write

$$\sigma(T_n) \cap \Omega_\varepsilon(\mu) = \{\mu_{nk} \colon k = 1, \ldots, k_n\}, \qquad \varepsilon = \varepsilon(\mu, T).$$

It follows that

$$\sigma(T_n) \cap \Omega_\varepsilon(\mu) = \{\mu_{nk}: k = 1, \ldots, k_n\} \quad \text{for} \quad n \geqq N_\varepsilon, \qquad \varepsilon \leqq \varepsilon(\mu, T).$$

Therefore,

$$\max_k |\mu_{nk} - \mu| \to 0 \qquad \text{as} \qquad n \to \infty.$$

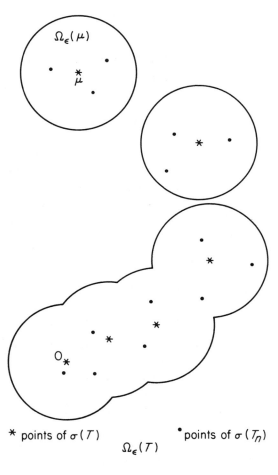

* points of $\sigma(T)$ •points of $\sigma(T_n)$

$$\Omega_\epsilon(T)$$

Figure 1

Next we examine eigenvectors and eigenmanifolds of the operators T and T_n. A comparison of dimensions of corresponding eigenmanifolds will be accomplished with the aid of the Riesz lemma.

66

LEMMA 4.9. Let $\mathscr{M}$ be a closed proper subspace of $\mathscr{X}$. For each $\varepsilon > 0$ there exists $x_\varepsilon \in \mathscr{X}$ such that

$$\|x_\varepsilon\| = 1, \qquad \|x_\varepsilon - y\| \geqq 1 - \varepsilon \quad \text{for} \quad y \in \mathscr{M}.$$

If dim $\mathscr{M} < \infty$, there exists $x \in \mathscr{X}$ such that

$$\|x\| = 1, \qquad \|x - y\| \geqq 1 \quad \text{for} \quad y \in \mathscr{M}.$$

COROLLARY 4.10. Let $\mathscr{M}$ be a subspace of $\mathscr{X}$ with dim $\mathscr{M} = m$. Then $\mathscr{M}$ has a basis $\{x_k : k = 1, \ldots, m\}$ such that

$$\|x_k\| = 1, \qquad \left\| x_k - \sum_{j=1}^{k-1} c_j x_j \right\| \geqq 1$$

for $k = 1, \ldots, m$ and any choices of the scalars c_j.

THEOREM 4.11. Let $T, T_n \in [\mathscr{X}]$, $n = 1, 2, \ldots$ Assume $T_n \to T$, $\{T_n\}$ collectively compact, and T compact. Let $\mu_n \in \sigma(T_n)$ and $\mu_n \to \mu \neq 0$. Then $\mu \in \sigma(T)$. If $\{x_n\}$ is bounded and $x_n \in \mathscr{N}(\mu_n - T_n)$, then there exist $\{T_{n_i}\}$, $\{x_{n_i}\}$ and $x \in \mathscr{X}$ such that $x_{n_i} \to x \in \mathscr{N}(\mu - T)$. For n sufficiently large,

$$\dim \mathscr{N}(\mu_n - T_n) \leqq \dim \mathscr{N}(\mu - T).$$

Let $\mathscr{M} \subset \mathscr{N}(\mu - T)$ and $\mathscr{M}_n \subset \mathscr{N}(\mu_n - T_n)$ be subspaces such that

$$x_n \in \mathscr{M}_n, \qquad x_n \to x \qquad \Rightarrow \qquad x \in \mathscr{M}.$$

Then dim $\mathscr{M}_n \leqq$ dim $\mathscr{M}$ eventually.

Proof. Without loss of generality, $\mu_n \neq 0$ for all n. Let $K = T/\mu$ and $K_n = T_n/\mu_n$ in the proof of Theorem 1.6 to obtain the assertions involving $x_{n_i} \to x \in \mathscr{N}(\mu - T)$. Hence, $\mu \in \sigma(T)$. Note that special cases of $\mathscr{M}$ and $\mathscr{M}_n$ are $\mathscr{M} = \mathscr{N}(\mu - T)$ and $\mathscr{M}_n = \mathscr{N}(\mu_n - T_n)$. Thus, it remains only to prove that dim $\mathscr{M}_n \leqq$ dim $\mathscr{M}$ eventually. Suppose that dim $\mathscr{M}_n \geqq m$ for all n in an infinite set J. Then there exist $x_{nk} \in \mathscr{M}_n$ such that

$$\|x_{nk}\| = 1, \qquad \left\| x_{nk} - \sum_{j=1}^{k-1} c_j x_j \right\| \geqq 1$$

for $n \in J$, $k = 1, \ldots, m$, and any choices of the c_j. By the hypotheses on $\mathcal{M}$ and $\mathcal{M}_n$, and the part of the theorem already proved, there exist $\{T_{n_i}\}$, $\{x_{n_i k}\}$ and $x_k \in \mathcal{X}$ such that $x_{n_i k} \to x_k \in \mathcal{M}$, $k = 1, \ldots, m$. It follows that

$$\|x_k\| = 1, \qquad \left\| x_k - \sum_{j=1}^{k-1} c_j x_j \right\| \geq 1$$

for $k = 1, \ldots, m$ and any choices of the c_j. Therefore, $\{x_1, \ldots, x_m\}$ is linearly independent and dim $\mathcal{M} \geq m$. Contrapositively, if dim $\mathcal{M} < m$, then dim $\mathcal{M}_n < m$ for all n sufficiently large. So dim $\mathcal{M}_n \leq$ dim $\mathcal{M}$ eventually and the theorem is proved.

In order to sharpen Theorem 4.11 in an important special case, the following elementary result is needed.

LEMMA 4.12. Let $T, T_n \in [\mathcal{X}]$, dim $T\mathcal{X} < \infty$, for all n sufficiently large.

Proof. Let $\{Tx_j : j = 1, \ldots, m\}$ be linearly independent and define

$$\mathcal{S} = \left\{ \sum_{j=1}^{m} c_j x_j : \max |c_j| = 1 \right\}.$$

Then $\mathcal{S}$ is compact, $T\mathcal{S}$ is compact, $T_n \to T$ uniformly on $\mathcal{S}$, and $\min_{x \in \mathcal{S}} \|Tx\| > 0$. Hence, for all n sufficiently large, $\min_{x \in \mathcal{S}} \|T_n x\| > 0$ and $\{T_n x_j : j = 1, \ldots, m\}$ is linearly independent. The desired result follows.

Next we specialize T and T_n in Theorem 4.11 to be projections. Recall that $E \in [\mathcal{X}]$ is a projection iff $E^2 = E$, in which case $I - E$ is the complementary projection,

$$E\mathcal{X} = \mathcal{N}(I - E), \qquad (I - E)\mathcal{X} = \mathcal{N}(E),$$

and

$$\mathcal{X} = E\mathcal{X} \oplus (I - E)\mathcal{X},$$

i.e., each $x \in \mathcal{X}$ has a unique representation $x = y + z$ with $y \in E\mathcal{X}$ and $z \in (I - E)\mathcal{X}$.

THEOREM 4.13. Let $E, E_n \in [\mathscr{X}]$, $n = 1, 2, \ldots$, be projections such that $E_n \to E$ and $\{E_n - E\}$ is collectively compact. Then $\dim E_n \mathscr{X} = \dim E \mathscr{X}$ for all n sufficiently large. (All infinite cardinals are identified.)

Proof. A projection is compact iff its range is finite dimensional. Hence, $\dim E_n \mathscr{X} = \dim E \mathscr{X} = +\infty$ if none of the projections E and E_n is compact. Assume now that E or some E_n is compact. By Proposition 4.1, $\{E_n\}$ is collectively compact and E is compact. Hence, $\dim E \mathscr{X} < \infty$ and $\dim E_n \mathscr{X} < \infty$ for all n. Since $E \mathscr{X} = \mathscr{N}(I - E)$ and $E_n \mathscr{X} = \mathscr{N}(I - E_n)$, Theorem 4.11 yields $\dim E_n \mathscr{X} \leq \dim E \mathscr{X}$ eventually. The opposite inequality follows from Lemma 4.12.

In §4.8, Theorem 4.13 will be applied to the situation with E a spectral projection and $E \mathscr{X}$ a spectral subspace. This will enable us to obtain a partial converse of Theorem 4.11: If $\mathscr{X}$ is complex and $\mu \in \sigma(T)$, $\mu \neq 0$, then there exist $\mu_n \in \sigma(T_n)$ such that $\mu_n \to \mu$.

4.6   GENERALIZED EIGENMANIFOLDS

The dimension inequality in Theorem 4.11 will be extended to generalized eigenmanifolds of T and T_n. We use the notation introduced in the first two paragraphs of §4.5.

THEOREM 4.14. Let $T, T_n \in [\mathscr{X}], n = 1, 2, \ldots$. Assume $T_n \to T$, $\{T_n\}$ collectively compact, and T compact. Fix $\mu \in \sigma(T)$, $\mu \neq 0$. Choose any nonnegative integers γ and γ_{nk}, $k = 1, \ldots, k_n$, such that

$$\sum_{k=1}^{k_n} \gamma_{nk} \leq \gamma.$$

Then, for all n sufficiently large,

$$\sum_{k=1}^{k_n} \dim \mathscr{N}[(\mu_{nk} - T_n)^{\gamma_{nk}}] \leq \dim \mathscr{N}[(\mu - T)^\gamma].$$

Proof. Without loss of generality,

$$\sum_{k=1}^{k_n} \gamma_{nk} = \gamma \quad \text{for all} \quad n.$$

69

From linear algebra [70, p. 317],

$$\mathcal{N}\left[\prod_{k=1}^{k_n}(\mu_{nk} - T_n)^{\gamma nk}\right] = \bigoplus_{k=1}^{k_n} \mathcal{N}[(\mu_{nk} - T_n)^{\gamma nk}],$$

$$\dim \mathcal{N}\left[\prod_{k=1}^{k_n}(\mu_{nk} - T_n)^{\gamma nk}\right] = \sum_{k=1}^{k_n}\dim \mathcal{N}[(\mu_{nk} - T_n)^{\gamma nk}].$$

Define μ_n, $\tilde{T}_n$ and $\tilde{T}$ by

$$\prod_{k=1}^{k_n}(\mu_{nk} - T_n)^{\gamma nk} = \mu_n - \tilde{T}_n, \qquad \mu_n = \prod_{k=1}^{k_n}(\mu_{nk})^{\gamma nk},$$

$$(\mu - T)^{\gamma} = \mu^{\gamma} - \tilde{T}.$$

Then $\tilde{T}_n \to \tilde{T}$ and $\mu_n \to \mu^{\gamma}$. By Proposition 4.2, $\tilde{T}$ is compact and $\{T_n\}$ is collectively compact. By Theorem 4.11, eventually

$$\dim \mathcal{N}(\mu_n - \tilde{T}_n) \leq \dim \mathcal{N}(\mu^{\gamma} - \tilde{T}).$$

The assertion follows.

Let $T \in [\mathcal{X}]$ be compact and $\mu \in \sigma(T)$, $\mu \neq 0$. Then there exists a positive integer $\nu = \nu(\mu - T)$ such that (cf. Appendix 1)

$$\mathcal{N}(\mu - T) \underset{\neq}{\subset} \cdots \underset{\neq}{\subset} \mathcal{N}[(\mu - T)^{\nu}] = \mathcal{N}[(\mu - T)^{\nu+1}] = \cdots.$$

The integer ν is called the *index* of $\mu - T$, or simply of μ. There is no loss in generality if $\gamma \leq \nu(\mu - T)$ and $\gamma_{nk} \leq \nu(\mu_{nk} - T_n)$ in Theorem 4.14. A generalization of this theorem and some related results will be given in §4.9 for the case with $\mathcal{X}$ complex.

4.7 FUNCTIONS OF OPERATORS

For the remainder of the chapter, $\mathcal{X}$ is a complex Banach space. We shall be concerned with functions of operators defined in terms of analytic functions of a complex variable. A short summary of fundamental definitions and properties will be given.

For each $T \in [\mathscr{X}]$, let $\mathscr{F}(T)$ denote the family of all complex functions f which are analytic on open, not necessarily connected, domains $\mathscr{D}(f) \supset \sigma(T)$. Given T and f, there is an open set Ω such that

$$\sigma(T) \subset \Omega \subset \bar{\Omega} \subset \mathscr{D}(f)$$

and the boundary Γ of Ω consists of a finite number of rectifiable Jordan curves. The operator $f(T) \in [\mathscr{X}]$ is defined by

$$f(T) = \frac{1}{2\pi i} \int_\Gamma f(\lambda)(\lambda - T)^{-1} \, d\lambda,$$

where Γ has positive orientation and the integral is the limit in operator norm of the usual approximating sums. Then

$$f(T)x = \frac{1}{2\pi i} \int_\Gamma f(\lambda)(\lambda - T)^{-1}x \, d\lambda, \qquad x \in \mathscr{X}.$$

The operator $f(T)$ is independent of the particular choice of the contour Γ. Hence, if $f,g \in \mathscr{F}(T)$ and $f(\lambda) = g(\lambda)$ on a neighborhood of $\sigma(T)$, then $f(T) = g(T)$.

Important examples are

$$f(T) = O \quad \text{when} \quad f(\lambda) \equiv 0,$$

$$f(T) = I \quad \text{when} \quad f(\lambda) \equiv 1,$$

$$f(T) = T \quad \text{when} \quad f(\lambda) \equiv \lambda.$$

If $f,g \in \mathscr{F}(T)$ and α, β are complex numbers, then $\alpha f + \beta g$ and fg are in $\mathscr{F}(T)$ and

$$(\alpha f + \beta g)(T) = \alpha f(T) + \beta g(T),$$

$$(fg)(T) = f(T)g(T).$$

Hence, if $f(\lambda)$ is any polynomial in λ, then $f(T)$ is the corresponding polynomial in T.

For derivations of these and other properties of functions of operators see [38] or [70].

THEOREM 4.15. Let $T, T_n \in [\mathscr{X}]$, $n = 1, 2, \ldots$. Assume $T_n \to T$ and $\{T_n - T\}$ collectively compact. Let $f \in \mathscr{F}(T)$. Then there exists N

such that

$$f \in \mathscr{F}(T_n) \quad \text{for} \quad n \geq N,$$
$$f(T_n) \to f(T),$$
$$\{f(T_n) - f(T): n \geq N\} \text{ is collectively compact.}$$

Proof. Let Γ be an admissible contour for $f(T)$. By Theorem 4.8 there exists N such that, for each $n \geq N$, $f \in \mathscr{F}(T_n)$ and Γ is an admissible contour for $f(T_n)$. Then

$$f(T_n) - f(T) = \frac{1}{2\pi i} \int_\Gamma f(\lambda)[(\lambda - T_n)^{-1} - (\lambda - T)^{-1}] \, d\lambda, \qquad n \geq N.$$

The remaining assertions now follow from Theorem 4.7 and Proposition 4.2.

Fix $\mu \in \rho(T)$ and define $f(\lambda) = (\mu - \lambda)^{-1}$ for $\lambda \neq \mu$. Then $f \in \mathscr{F}(T)$ and $f(T) = (\mu - T)^{-1}$. By Theorem 4.15, there exists N such that

$$\mu \in \rho(T_n) \quad \text{for} \quad n \geq N,$$
$$(\mu - T_n)^{-1} \to (\mu - T)^{-1},$$
$$\{(\mu - T_n)^{-1} - (\mu - T)^{-1}: n \geq N\} \text{ is collectively compact.}$$

These results were obtained more directly in Theorem 4.3.

In the next section an application of Theorem 4.15 in which $f(T)$ and $f(T_n)$ are spectral projections will be given.

4.8 SPECTRAL PROJECTIONS AND SPECTRAL SUBSPACES

A *spectral set* for an operator $T \in [\mathscr{X}]$ is closed subset σ of $\sigma(T)$ such that $\sigma(T) - \sigma$ also is closed. Given T and σ, there exists $e \in \mathscr{F}(T)$ such that $e(\lambda) \equiv 1$ on σ and $e(\lambda) \equiv 0$ on $\sigma(T) - \sigma$. Let

$$E = E(\sigma, T) = e(T).$$

Then

$$E = \frac{1}{2\pi i} \int_\Gamma (\lambda - T)^{-1} \, d\lambda$$

for any contour Γ with σ inside and $\sigma(T) - \sigma$ outside. Since $e^2 = e$, we have $E^2 = E$. Thus, E is a projection. It is the *spectral projection* associated with σ and T. Its range $E\mathscr{X}$ is the *spectral subspace* associated with σ and T. In particular,

$$E = O \text{ iff } \sigma = \phi,$$

$$E = I \text{ iff } \sigma = \sigma(T).$$

The complement $\sigma' = \sigma(T) - \sigma$ is a spectral set with associated spectral projection $E' = I - E$. Thus, E and E' are complementary projections and

$$\mathscr{X} = E\mathscr{X} \oplus E'\mathscr{X}.$$

The significance of spectral subspaces is indicated by the following important facts. First, $T(E\mathscr{X}) \subset E\mathscr{X}$. Let $T_{E\mathscr{X}}$ denote the restriction of T to $E\mathscr{X}$. Then

$$\sigma(T_{E\mathscr{X}}) = \sigma, \qquad \sigma_p(T_{E\mathscr{X}}) = \sigma \cap \sigma_p(T),$$

and a set σ_0 in the complex plane is a spectral set for $T_{E\mathscr{X}}$ iff σ_0 is a spectral set for T such that $\sigma_0 \subset \sigma$, in which case the associated spectral subspaces coincide:

$$E(\sigma_0, T_{E\mathscr{X}})(E\mathscr{X}) = E(\sigma_0, T)\mathscr{X}.$$

If $\dim E\mathscr{X} < \infty$, then $\sigma \subset \sigma_p(T)$ and $E\mathscr{X}$ is spanned by the generalized eigenvectors corresponding to the eigenvalues of T in σ.

The final theorem of this section compares spectral projections of operators T and T_n which satisfy the hypotheses in §4.1. It is a direct consequence of Theorems 4.15 and 4.13.

THEOREM 4.16. Let $T, T_n \in [\mathscr{X}]$, $n = 1, 2, \ldots$. Assume $T_n \to T$ and $\{T_n - T\}$ collectively compact. Let σ be a spectral set for T, $E = E(\sigma, T)$, and Γ a contour with σ inside and $\sigma(T) - \sigma$ outside. Then there exists N such that for $n \geq N$:

$\Gamma \subset \rho(T_n)$,

$\sigma_n = \{\lambda \in \sigma(T_n) : \lambda \text{ inside } \Gamma\}$ is a spectral set for T_n,

the spectral projection $E_n = E(\sigma_n, T_n)$ is defined,

$E_n \to E$,

$\{E_n - E\}$ is collectively compact,

$\dim E_n\mathscr{X} = \dim E\mathscr{X}$ (finite or infinite),

$\sigma_n = \phi$ iff $\sigma = \phi$.

4.9 EIGENVALUES AND
GENERALIZED EIGENMANIFOLDS REVISITED

Throughout this section the hypotheses and notation of Theorem 4.16 will prevail. Without essential loss of generality, $N = 1$. These assumptions will not be repeated in statements of theorems.

The easiest case is when σ consists of a simple eigenvalue μ.

THEOREM 4.17. Let $\sigma = \{\mu\}$ and dim $E\mathcal{X} = 1$. Then there is a non-zero $x \in \mathcal{X}$ such that

$$Tx = \mu x, \qquad Ex = x, \qquad E\mathcal{X} = \operatorname{span}\{x\}.$$

For each $n \geq 1$ there exists μ_n such that

$$\sigma_n = \{\mu_n\}, \qquad \mu_n \to \mu.$$

Let $x_n = E_n x$. Then

$$T_n x_n = \mu_n x_n, \qquad x_n \to x.$$

For each n sufficiently large, $x_n \neq 0$ and $E_n\mathcal{X} = \operatorname{span}\{x_n\}$.

Proof. Apply Theorems 4.8 and 4.16.

Now we consider a more general situation. For some μ, let $\sigma = \{\mu\}$ and dim $E\mathcal{X} < \infty$. Then $\mu \in \sigma_p(T)$, the index $\nu = \nu(\mu - T)$ is finite, and

$$E\mathcal{X} = \mathcal{N}[(\mu - T)^\nu].$$

(In this case, μ is a pole of order ν for the resolvent of T.)

By Theorem 4.16, dim $E_n\mathcal{X} = \dim E\mathcal{X} < \infty$. So the spectral set σ_n consists of a finite number of eigenvalues μ_{nk}, $k = 1, \ldots, k_n$, of T_n with finite indices ν_{nk} and

$$E_n\mathcal{X} = \bigoplus_{k=1}^{k_n} \mathcal{N}[(\mu_{nk} - T_n)^{\nu_{nk}}] = \mathcal{N}\left[\prod_{k=1}^{k_n}(\mu_{nk} - T_n)^{\nu_{nk}}\right].$$

By Theorem 4.8,

$$\max_k |\mu_{nk} - \mu| \to 0 \quad \text{as} \quad n \to \infty.$$

For convenience, define polynomials

$$p(\lambda) = (\mu - \lambda)^\nu, \qquad p_n(\lambda) = \prod_{k=1}^{k_n} (\mu_{nk} - \lambda)^{\nu_{nk}}.$$

Then

$$p(T) = (\mu - T)^\nu, \qquad p_n(T_n) = \prod_{k=1}^{k_n} (\mu_{nk} - T_n)^{\nu_{nk}},$$

$$E\mathcal{X} = \mathcal{N}[p(T)], \qquad E_n\mathcal{X} = \mathcal{N}[p_n(T_n)].$$

$$Ep(T) = p(T)E = O, \qquad E_n p_n(T_n) = p_n(T_n)E_n = O.$$

Among all polynomials which satisfy the last two equations, p and p_n have minimal degrees. The degree of p_n is

$$\nu_n = \sum_{k=1}^{k_n} \nu_{nk}.$$

THEOREM 4.18. Eventually, $\nu_n \geqq \nu$.

Proof. Suppose that $\nu_n = \alpha$ for all n in an infinite set J. Since $T_n \to T$ and $E_n \to E$, it follows that, as $n \to \infty$ through J,

$$p_n(T_n) \to (\mu - T)^\alpha,$$

$$O = p_n(T_n)E_n \to (\mu - T)^\alpha E,$$

So $(\mu - T)^\alpha E = O$ and $\alpha \geqq \nu$. The assertion follows.

The next result compares the dimensions of generalized eigenmanifolds of T and T_n. It extends Theorem 4.14.

THEOREM 4.19. Choose any nonnegative integers γ and γ_{nk}, $k = 1, \ldots, k_n$, such that

$$\sum_{k=1}^{k_n} \gamma_{nk} \leqq \gamma.$$

Then

$$\sum_{k=1}^{k_n} \dim \mathcal{N}[(\mu_{nk} - T_n)^{\gamma_{nk}}] \leqq \dim \mathcal{N}[(\mu - T)^\gamma]$$

eventually.

Proof. Without loss of generality,

$$\sum_{k=1}^{k_n} \gamma_{nk} = \gamma \quad \text{for all} \quad n.$$

Let

$$q(T) = (\mu - T)^\gamma, \qquad q_n(T_n) = \prod_{k=1}^{k_n} (\mu_{nk} - T_n)^{\gamma_{nk}}.$$

Note that

$$\mathcal{N}[q(T)] \subset E\mathcal{X} = \mathcal{N}(I - E),$$

$$\mathcal{N}[q_n(T_n)] \subset E_n\mathcal{X} = \mathcal{N}(I - E_n),$$

$$q_n(T_n) \rightarrow q(T).$$

Therefore,

$$x_n \in \mathcal{N}[q_n(T_n)], \quad x_n \rightarrow x \quad \Rightarrow \quad x \in \mathcal{N}[q(T)].$$

By Theorem 4.11 with $\mu - T$ and $\mu_n - T_n$ replaced by $I - E$ and $I - E_n$, $\dim \mathcal{N}[q_n(T_n)] \leqq \dim \mathcal{N}[q(T)]$ eventually. Since

$$\mathcal{N}[q_n(T_n)] = \bigoplus_{k=1}^{k_n} \mathcal{N}[(\mu_{nk} - T_n)^{\gamma_{nk}}],$$

the desired inequality follows.

Suppose $0 \leqq \gamma \leqq \nu$ in Theorem 4.19. By Theorem 4.18, for n sufficiently large there exist γ_{nk} such that

$$0 \leqq \gamma_{nk} \leqq \nu_{nk}, \qquad \sum_{k=1}^{k_n} \gamma_{nk} = \gamma.$$

Two particular cases of Theorem 4.19 are worth special notice. When $\gamma = 1$,

$$\dim \mathcal{N}(\mu_{nk} - T_n) \leqq \dim \mathcal{N}(\mu - T)$$

for all n sufficiently large and for all k. If T is compact this also follows from

Theorem 4.11. When $\gamma = \nu$ the conclusion of Theorem 4.19 is also a consequence of

$$\mathscr{N}[q(T)] = E\mathscr{X}, \qquad \mathscr{N}[q_n(T_n)] \subset E_n\mathscr{X},$$

$$\dim E_n\mathscr{X} = \dim E\mathscr{X}.$$

Incidentally, another proof of Theorem 4.18 is based on Theorem 4.19 with $\gamma = \nu$.

4.10 ERROR ESTIMATES FOR EIGENVECTORS

In the foregoing analysis, the spectral hypotheses were mainly on T and the conclusions on T_n. Now the roles of T and T_n are reversed. This material extends work of Atkinson [21].

THEOREM 4.20. Let $T, T_n \in [\mathscr{X}]$, $n = 1, 2, \ldots$. Assume $T_n \to T$ and $\{T_n - T\}$ collectively compact. Let E be a spectral projection for T with $\dim E\mathscr{X} < \infty$. Let E_n, $n \geq N$, be related spectral projections for T_n as defined in Theorem 4.16. Let

$$x_n \in E_n\mathscr{X}, \qquad \|x_n\| = 1 \quad \text{for} \quad n \geq N.$$

Then

$$\|x_n - Ex_n\| \to 0,$$

$$\|Ex_n\| \to 1,$$

$$Ex_n \neq 0 \text{ eventually}.$$

Proof. Since $\dim E\mathscr{X} < \infty$, E is compact. By Proposition 4.1 and Theorem 4.16, $\{E_n\}$ is collectively compact. Since $E_n x_n = x_n$ and $x_n - Ex_n = (E_n - E)E_n x_n$, Corollary 1.9 yields

$$\|x_n - Ex_n\| \leq \|(E_n - E)E_n\| \to 0.$$

Since $\|x_n\| = 1$ for $n \geq N$, the other conclusions follow.

Observe that $Ex_n \in E\mathscr{X}$ in Theorem 4.20. If $\dim E\mathscr{X} < \infty$, then Ex_n is a linear combination of eigenvectors or generalized eigenvectors of T.

THEOREM 4.21. Let $T, T_n \in [\mathcal{X}]$, $n = 1, 2, \ldots$. Assume $T_n \to T$ and $\{T_n - T\}$ collectively compact. Let σ be a spectral set for T and $E = E(\sigma, T)$. Define σ_n and $E_n = E(\sigma_n, T_n)$ for $n \geq N$ as in Theorem 4.16. Suppose

$$T_n x_n = \mu_n x_n, \qquad \mu_n \in \sigma_n, \qquad \|x_n\| = 1 \quad \text{for} \quad n \geq N.$$

Then

$$\|x_n - E x_n\| \leq r_n$$

where

$$r_n = \frac{\ell(\Gamma)}{2\pi} \|\mu_n x_n - T x_n\| \max_{\lambda \in \Gamma} \frac{\|(\lambda - T)^{-1}\|}{|\lambda - \mu_n|}.$$

If, for some n, $r_n < 1$, then $E x_n \neq 0$ and $\sigma \neq \phi$. If $\dim E\mathcal{X} < \infty$, then $r_n \to 0$.

Proof. Note that $E_n x_n = x_n$,

$$E_n - E = \frac{1}{2\pi i} \int_\Gamma (\lambda - T)^{-1} (T_n - T)(\lambda - T_n)^{-1} \, d\lambda,$$

and $(\lambda - T_n)^{-1} x_n = (\lambda - \mu_n)^{-1} x_n$ for $\lambda \in \rho(T_n)$. Therefore,

$$x_n - E x_n = \frac{1}{2\pi i} (\mu_n x_n - T x_n) \int_\Gamma \frac{(\lambda - T)^{-1}}{\lambda - \mu_n} \, d\lambda,$$

which implies $\|x_n - E x_n\| \leq r_n$. Since $\mu_n \notin \Gamma$,

$$\delta_n = \min_{\lambda \in \Gamma} |\lambda - \mu_n| > 0$$

and

$$r_n \leq \frac{\ell(\Gamma)}{2\pi \delta_n} \|\mu_n x_n - T x_n\| \max_{\lambda \in \Gamma} \|(\lambda - T)^{-1}\|.$$

By Theorems 4.8 and 4.16, there exists δ such that $\delta_n \geq \delta > 0$ for all n sufficiently large. Now

$$\|\mu_n x_n - T x_n\| \leq \|(T_n - T)E_n\|,$$

$$r_n \leq \frac{\ell(\Gamma)}{2\pi \delta} \|(T_n - T)E_n\| \max_{\lambda \in \Gamma} \|(\lambda - T)^{-1}\|.$$

If $\dim E\mathcal{X} < \infty$, then E is compact, $\{E_n\}$ is collectively compact, and $\|(T_n - T)E_n\| \to 0$, so that $r_n \to 0$.

Other estimates for r_n can be obtained by using the inequality for $\|(\lambda - T)^{-1}\|$ in terms of $\|(\lambda - T_n)^{-1}\|$ which appears in Theorem 4.6.

Theorem 4.21 provides the criterion $r_n < 1$, based on a single value of n, for the spectral set σ to be nonvoid. However, this is of limited practical value because $\|(\lambda - T)^{-1}\|$ becomes large and $|\lambda - \mu_n|$ may become small if a point of $\sigma(T)$ is near the contour Γ. For a further discussion of this topic, see Atkinson [21].

chapter **5**

CHARACTERIZATIONS OF COLLECTIVELY COMPACT AND TOTALLY BOUNDED SETS OF OPERATORS

5.I INTRODUCTION

As before, let $\mathscr{X}$ be a real or complex Banach space, $\mathscr{B}$ the closed unit ball in $\mathscr{X}$, and $[\mathscr{X}]$ the space of bounded linear operators on $\mathscr{X}$ equipped with the operator norm topology.

In this chapter, collectively compact sets of operators in $[\mathscr{X}]$ are compared with bounded sets and also with totally bounded sets of compact operators in $[\mathscr{X}]$. The results yield information on the scope and applicability of the approximation theory developed in Chapters 1 and 4. The principal theorem states that a set $\mathscr{K}$ of compact operators in $[\mathscr{X}]$ is totally bounded iff both $\mathscr{K}$ and the set $\{K^*: K \in \mathscr{K}\}$ of adjoint operators are collectively compact.

References are [9, 10, 15, 17, 65] by Anselone and Palmer. Most of the results in these papers pertain to the more general situation in which the operators map one normed linear space, not necessarily complete, to another.

81

5.2 BOUNDED SETS OF COMPACT OPERATORS

A set $\mathcal{K} \subset [\mathcal{X}]$ is bounded iff $\mathcal{K}\mathcal{B}$ is bounded, whereas $\mathcal{K}$ is collectively compact iff $\mathcal{K}\mathcal{B}$ is totally bounded. Therefore, every collectively compact set $\mathcal{K} \subset [\mathcal{X}]$ is a bounded set of compact operators. The converse is false.

EXAMPLE 5.1. Let $\mathcal{X}$ be an infinite dimensional Hilbert space (e.g., ℓ^2) with orthonormal basis $\{\varphi_\alpha : \alpha \in A\}$. Let E_α be the orthogonal projection onto the one-dimensional subspace spanned by φ_α. Then

$$E_\alpha x = (x, \varphi_\alpha)\varphi_\alpha, \qquad \|E_\alpha\| = 1,$$

and E_α is compact since $\dim (E_\alpha \mathcal{X}) < \infty$. Thus, $\mathcal{E} = \{E_\alpha : \alpha \in A\}$ is a bounded set of compact operators in $[\mathcal{X}]$. However, $\mathcal{E}$ is not collectively compact because

$$\varphi_\alpha = E_\alpha \varphi_\alpha \in \mathcal{E}\mathcal{B} \quad \text{for} \quad \alpha \in A,$$

$$\|\varphi_\alpha - \varphi_\beta\| = \sqrt{2} \quad \text{for} \quad \alpha \neq \beta, \qquad \alpha, \beta \in A,$$

which imply that $\mathcal{E}\mathcal{B}$ is not totally bounded.

5.3 COLLECTIVELY COMPACT AND TOTALLY BOUNDED SETS

The approximation theory developed in previous chapters is based on pointwise operator convergence. Since there is an alternative approximation theory for operator norm convergence [49, 67], it is important to distinguish between the two types of convergence.

LEMMA 5.2. Let $T, T_n \in [\mathcal{X}]$, $n = 1, 2, \ldots$. Then $\|T_n - T\| \to 0$ iff $T_n \to T$ and $\{T_n\}$ is totally bounded (equivalently, relatively compact or sequentially compact) with respect to the norm topology on $[\mathcal{X}]$.

Proof. The forward assertion is clear enough. The converse is a consequence of the fact that, in a metric space, $z_n \to z$ whenever every subsequence of $\{z_n\}$ has a further subsequence which converges to z.

In view of Lemma 5.2, the approximation theory of Chapter 1 is most appropriate for operators $K, K_n \in [\mathscr{X}]$ such that $K_n \to K$ and $\{K_n\}$ is collectively compact but not totally bounded. This is the case for the applications to integral equations in Chapters 2 and 3. The approximation theory in Chapter 4 pertains especially to operators $T, T_n \in [\mathscr{X}]$ such that $T_n \to T$ and $\{T_n - T\}$ is collectively compact but not totally bounded.

The foregoing discussion motivates a comparison between collectively compact and totally bounded sets of operators.

PROPOSITON 5.3. Let $\mathscr{K}$ be a totally bounded set of compact operators in $[\mathscr{X}]$. Then $\mathscr{K}$ is collectively compact.

Proof. Fix $\varepsilon > 0$. Then $\mathscr{K}$ has a finite ε-net $\mathscr{K}_\varepsilon \subset \mathscr{K}$: for each $K \in \mathscr{K}$ there exists $K_\varepsilon \in \mathscr{K}_\varepsilon$ such that

$$\|Kx - K_\varepsilon x\| \leq \|K - K_\varepsilon\| < \varepsilon \quad \text{for} \quad x \in \mathscr{B}.$$

Hence, $\mathscr{K}\mathscr{B}$ has the ε-net $\mathscr{K}_\varepsilon \mathscr{B}$ which is totally bounded because $\mathscr{K}_\varepsilon$ is a finite set of compact operators. It follows that $\mathscr{K}\mathscr{B}$ is totally bounded and $\mathscr{K}$ is collectively compact.

The converse of Proposition 5.3 is false. Counterexamples are given by the integral equations applications and by the following more elementary illustration.

EXAMPLE 5.4. Let $\mathscr{X} = \ell^2$. Define $K_n \in [\mathscr{X}]$, $n = 1, 2, \ldots$, by

$$K_n x = x_n \varphi_1, \qquad x = (x_1, x_2, \ldots, x_n \ldots),$$

where $\varphi_1 = (1, 0, 0, \ldots)$. Let $\mathscr{K} = \{K_n\}$. Then $\mathscr{K}\mathscr{B}$ is bounded and dim $\mathscr{K}\mathscr{X} = 1$, so that $\mathscr{K}$ is collectively compact. But $\mathscr{K}$ is not totally bounded, for

$$\|K_m - K_n\| = \sqrt{2}, \qquad m \neq n.$$

Partial converses of Proposition 5.3, involving adjoint operators, will be given.

5.4 COLLECTIVELY COMPACT SETS OF OPERATORS AND THEIR ADJOINTS

Let $\mathscr{X}^*$ be the normed dual of $\mathscr{X}$. The *adjoint* of $K \in [\mathscr{X}]$ is the unique operator $K^* \in [\mathscr{X}^*]$ such that

$$(K^*f)(x) = f(Kx) \quad \text{for} \quad x \in \mathscr{X}, \qquad f \in \mathscr{X}^*.$$

When $\mathscr{X}$ is a Hilbert space, it is customary and more convenient for K^* to denote the usual *Hilbert space adjoint* of $K \in [\mathscr{X}]$. Then $K^* \in [\mathscr{X}]$ and

$$(Kx, y) = (x, K^*y) \quad \text{for} \quad x, y \in \mathscr{X}.$$

For the rest of the chapter, statements concerning adjoint operators will have two equally valid interpretations when $\mathscr{X}$ is a Hilbert space.

For each $\mathscr{K} \subset [\mathscr{X}]$, define $\mathscr{K}^* = \{K^* : K \in \mathscr{K}\}$. Since $\|K^*\| = \|K\|$ for $K \in [\mathscr{X}]$, we have

$$\mathscr{K} \text{ totally bounded} \Leftrightarrow \mathscr{K}^* \text{ totally bounded}.$$

For each $K \in [\mathscr{X}]$, K is compact iff K^* is compact [70, pg. 485]. However,

$$\mathscr{K} \text{ collectively compact} \nLeftrightarrow \mathscr{K}^* \text{ collectively compact}.$$

Example 5.4 furnishes a collectively compact set $\mathscr{K}$ for which $\mathscr{K}^*$ is not collectively compact. This is rather easy to verify directly. It is also a consequence of the following theorem.

THEOREM 5.5 Let $\mathscr{K}$ be a set of compact operators in $[\mathscr{X}]$. Then $\mathscr{K}$ is totally bounded iff both $\mathscr{K}$ and $\mathscr{K}^*$ are collectively compact.

Proposition 5.3 yields the forward assertion. The converse, which is more difficult, has had an interesting history. First, Anselone and Palmer

[15] used Hilbert space spectral theory to establish the following theorem for normal operators $(KK^* = K^*K)$ on a Hilbert space.

THEOREM 5.6. Let $\mathscr{X}$ be a Hilbert space and $\mathscr{K}$ a set of compact normal operators on $[\mathscr{X}]$. Then

$$\mathscr{K} \quad \text{is totally bounded}$$

$$\Leftrightarrow \quad \mathscr{K} \quad \text{is collectively compact}$$

$$\Leftrightarrow \quad \mathscr{K}^* \text{ is collectively compact.}$$

If $\mathscr{K}$ and $\mathscr{K}^*$ are collectively compact sets in $[\mathscr{X}]$, with $\mathscr{X}$ a Hilbert space, then

$$\{K + K^*\colon K \in \mathscr{K}\}, \qquad \{K - K^*\colon K \in \mathscr{K}\}$$

are collectively compact sets of normal operators. By Theorem 5.6, these sets are totally bounded. Since $K = \frac{1}{2}\,[(K + K^*) + (K - K^*)]$, $\mathscr{K}$ is totally bounded. Thus, Theorem 5.5 is valid when $\mathscr{X}$ is a Hilbert space. Either Theorem 5.5 or 5.6 yields the following corollary.

COROLLARY 5.7. Let $\mathscr{X}$ be a Hilbert space and $\mathscr{K}$ a set of compact self-adjoint operators in $[\mathscr{X}]$. Then $\mathscr{K}$ is totally bounded iff $\mathscr{K}$ is collectively compact.

In view of the discussion in §5.2, the approximation methods in this book are particularly applicable to collectively compact sequences of operators for which the sequences of adjoint operators are not collectively compact. When $\mathscr{X}$ is a Hilbert space, this would mean that the operators fail to be self-adjoint or even normal.

An effort was made to extend the Hilbert space analysis as much as possible to the case with $\mathscr{X}$ a Banach space. As part of that program, Anselone [9] proved Theorem 5.5 for sets of operators with uniformly bounded (finite) ranks. Several side results on operator approximations and on the estimation of linear independence of vectors were derived in the process. The main conclusions of that paper were applied along with spectral theory by Anselone and Palmer [16] to obtain a proof of Theorem 5.5 for sets of normal operators on a complex, uniformly smooth Banach space $\mathscr{X}$. A number of other results were derived. It was shown that several spectral properties of compact operators hold uniformly over a collectively compact set. Collectively compact and totally bounded sets of spectral projections were studied as well.

Later, Palmer [65] managed to obtain the following theorem (in fact, for operators from one normed linear space to another) which implies the converse assertion in Theorem 5.5.

THEOREM 5.8. Let $\mathscr{K} \subset [\mathscr{X}]$. Assume
  (a) $\mathscr{K}\mathscr{B}$ is totally bounded in $\mathscr{X}$,
  (b) $\mathscr{K}^*f$ is totally bounded in $\mathscr{X}^*$ for each $f \in \mathscr{X}^*$.
Then $\mathscr{K}$ is totally bounded in $[\mathscr{X}]$.

Palmer's proof does not involve spectral theory. His reasoning is similar to that used by Vala [72] in a proof of a generalization of the Arzelà-Ascoli theorem. Later, J. Lindenstrauss (private communication) used sequential compactness and certain norm inequalities involving operators and their adjoints to obtain another proof, by contradiction, of Theorem 5.8. His arguments were substantially modified by Anselone [10] to produce a direct proof, given in §5.6. It is based on a new criterion for a set in a metric space to be totally bounded.

5.5 TOTALLY BOUNDED SETS IN METRIC SPACES

Recall that a set $\mathscr{S}$ in a metric space is totally bounded iff every sequence in $\mathscr{S}$ has a Cauchy subsequence. The following result is similar but has the advantage that sequences and subsequences are replaced by sets and subsets, so that multiple subscripts are avoided.

LEMMA 5.9. A set $\mathscr{S}$ in the metric space is totally bounded iff for each $\varepsilon > 0$ and each infinite set $\hat{\mathscr{S}} \subset \mathscr{S}$ there is an infinite set $\mathscr{S}_\varepsilon \subset \hat{\mathscr{S}}$ with diameter $(\mathscr{S}_\varepsilon) < \varepsilon$ (or $\leq \varepsilon$).

86

Proof. First let $\mathscr{S}$ be totally bounded. Fix $\varepsilon > 0$ and suppose that $\hat{\mathscr{S}}$ is an infinite subset of $\mathscr{S}$. Then $\hat{\mathscr{S}}$ is totally bounded. So it is a finite union of sets each with diameter less than ε. One of these sets, say $\mathscr{S}_\varepsilon$, must be infinite.

For the converse, assume that $\mathscr{S}$ is not totally bounded. Then there exist $\varepsilon > 0$ and an infinite set $\hat{\mathscr{S}} \subset \mathscr{S}$ such that

$$\text{dist}\,(x,y) \geqq \varepsilon \quad \text{for} \quad x \neq y, \quad x,y \in \hat{\mathscr{S}}.$$

Clearly, $\hat{\mathscr{S}}$ has no infinite subset with diameter less than ε.

Another form of the criterion in Lemma 5.9 is that for each infinite set $\hat{\mathscr{S}} \subset \mathscr{S}$ there exist infinite sets $\mathscr{S}_n \subset \hat{\mathscr{S}}$, $n = 1, 2, \ldots$, such that $\text{diam}\,(\mathscr{S}_n) \to 0$ as $n \to \infty$. If desired, $\mathscr{S}_{n+1} \subset \mathscr{S}_n$ for all n.

5.6 PROOFS OF THEOREM 5.8

First, let $\mathscr{X}$ be a Hilbert space. The reasoning is more transparent in this case and will serve to motivate the more general considerations which follow. Since Hilbert space adjoints will be used, condition (b) in Theorem 5.8 should be altered temporarily so that $f \in \mathscr{X}$. The proof will be based on Lemma 5.9 and on the Hilbert space identity

$$\|T\|^2 = \|T^*T\| \quad \text{for} \quad T \in [\mathscr{X}].$$

Proof 1. $\mathscr{X}$ **a Hilbert space.** Fix $\varepsilon > 0$ and suppose that $\hat{\mathscr{X}}$ is an infinite subset of $\mathscr{X}$. By hypothesis (a) of the theorem, $\mathscr{X}$ is bounded: $\|K\| \leqq b$ for all $K \in \mathscr{X}$ and some $b < \infty$. Let $\delta = \varepsilon^2/(2b + 1)$. Also by (a), the set

$$\mathscr{S} = \{Kx - Lx \colon K, L \in \mathscr{X}, \quad x \in \mathscr{B}\}$$

is totally bounded. Let $\mathscr{S}_\delta$ be a finite δ-net for $\mathscr{S}$. By hypothesis (b) and Lemma 5.9, there is an infinite set $\mathscr{X}_\varepsilon \subset \hat{\mathscr{X}}$ such that

$$\|(K - L)^* f\| < \delta \quad \text{for} \quad K, L \in \mathscr{X}_\varepsilon, \quad f \in \mathscr{S}_\delta.$$

A triangle inequality argument yields

$$\|(K - L)^* f\| < \varepsilon^2 \quad \text{for} \quad K, L \in \mathcal{K}_\varepsilon, \qquad f \in \mathcal{S}.$$

Hence, from the definition of $\mathcal{S}$,

$$\|K - L\|^2 = \|(K - L)^*(K - L)\| \leq \varepsilon^2 \quad \text{for} \quad K, L \in \mathcal{K}_\varepsilon.$$

By Lemma 5.9, $\mathcal{K}$ is totally bounded.

It is possible to carry over the foregoing proof almost without change to certain Banach spaces. We shall need the following concept. A linear functional $x^* \in \mathcal{X}^*$ is *conjugate* to $x \in \mathcal{X}$ provided that

$$\|x^*\| = \|x\|, \qquad x^*(x) = \|x\|^2.$$

For $\mathcal{X}$ a Hilbert space, each $x \in \mathcal{X}$ has a unique conjugate element $x^* \in \mathcal{X}^*$. It is determined by $x^*(y) = (y, x)$ for all $y \in \mathcal{X}$. This follows from the Riesz representation theorem and the condition for equality in the Schwarz inequality. For $\mathcal{X}$ a Banach space, the Hahn-Banach theorem implies that each $x \in \mathcal{X}$ has at least one conjugate element. For each $x \in \mathcal{X}$ let x^c denote the (nonvoid) set of elements $x^* \in \mathcal{X}^*$ which are conjugate to x.

The next result is a Banach space analogue of the Hilbert space identity $\|T\|^2 = \|T^*T\|$.

LEMMA 5.10. Let $T \in [\mathcal{X}]$. For each $x \in \mathcal{B}$ choose any $(Tx)^* \in (Tx)^c$. Then

$$\|T\|^2 = \sup_{x \in \mathcal{B}} \|T^*(Tx)^*\|.$$

Proof. For each $x \in \mathcal{B}$ and any $(Tx)^* \in (Tx)^c$,

$$\|Tx\|^2 = (Tx)^*(Tx) = (T^*(Tx)^*)x \leq \|T^*(Tx)^*\|,$$

$$\|T^*(Tx)^*\| \leq \|T^*\| \, \|(Tx)^*\| = \|T\| \, \|Tx\| \leq \|T\|^2.$$

The assertion follows.

Suppose temporarily that $\mathcal{X}$ is a Banach space with the special property: *For each totally bounded set $\mathcal{S} \subset \mathcal{X}$ there is a totally bounded set $\mathcal{S}^* = \{x^* : x \in \mathcal{S}\}$ with $x^* \in x^c$ for each $x \in \mathcal{S}$.* To adapt the Hilbert space proof of Theorem 5.8 to such a Banach space, define the totally bounded set $\mathcal{S}$ as before and choose a corresponding totally bounded set $\mathcal{S}^* = \{x^* : x \in \mathcal{S}\}$ of conjugate elements. Then for any $\varepsilon > 0$ and any infinite set $\hat{\mathcal{K}} \subset \mathcal{K}$, there is an infinite set $\mathcal{K}_\varepsilon \subset \hat{\mathcal{K}}$ such that

$$\|(K - L)^*(Kx - Lx)^*\| < \varepsilon^2 \quad \text{for} \quad K, L \in \mathcal{K}_\varepsilon, \qquad x \in \mathcal{B}.$$

By Lemmas 5.9 and 5.10, $\mathcal{K}$ is totally bounded.

Any Hilbert space has the special property described above. So does any Banach space such that each $x \in \mathcal{X}$ has a unique conjugate element $x^* \in \mathcal{X}^*$ and the map $x \mapsto x^*$ is continuous. This is the case, for example, if $\mathcal{X}^*$ is *uniformly rotund* (= uniformly convex), i.e., for each ε with $0 < \varepsilon < 2$ there exists $\delta > 0$ such that $\|f + g\| \leq 2(1 - \delta)$ whenever $f, g \in \mathcal{X}^*$, $\|f\| \leq 1$, $\|g\| \leq 1$, and $\|f - g\| > \varepsilon$. A proof that uniform rotundity implies the special property is given in [16].

Klee [50] has related the special property to the smoothness of the unit ball $\mathcal{B} \subset \mathcal{X}$. Among other results, he has shown that every infinite dimensional Banach space can be renormed so as to lack the property. Stanley Weiss (University of Chicago Ph.D. thesis, 1969) has proved that $\mathcal{X}$ has the property if the norm on $\mathcal{X}^*$ is locally uniformly convex; in particular, $\mathcal{X}^*$ has such an equivalent norm if it is separable. On the other hand, it is not known whether every Banach space (or even every separable Banach space) can be renormed so as to attain the property. If this were true, then the proof of Theorem 5.8, when adapted to a Banach space, would be valid in general. Since this question is unsettled, we shall modify the foregoing arguments so that the special property is no longer needed.

Again let $\mathcal{X}$ be a Banach space. The role of the special property now will be played by the following lemma.

LEMMA 5.11. Let $\mathcal{S} \subset \mathcal{X}$ be totally bounded and $\varepsilon > 0$. Then for each $x \in \mathcal{S}$ there exists $f_x \in \mathcal{X}^*$ such that

$$\big| \|f_x\| - \|x\| \big| < \varepsilon, \qquad \big| f_x(x) - \|x\|^2 \big| < \varepsilon.$$

$\mathcal{F} = \{f_x : x \in \mathcal{S}\}$ is finite.

Proof. Let b be a bound for $\mathcal{S}$ and $\delta = \min(\varepsilon, \varepsilon/3b)$. Let $\mathcal{S}_\delta \subset \mathcal{S}$ be a finite δ-net for $\mathcal{S}$. For each $x \in \mathcal{S}$ choose any $x_\delta \in \mathcal{S}_\delta$ such that $\|x - x_\delta\| < \delta$ and choose any $x_\delta^* \in x_\delta^c$. Define $f_x = x_\delta^*$. A triangle inequality argument completes the proof.

In Lemma 5.11, f_x may be called an ε-approximate conjugate element for x.

The previous roles of the Hilbert space identity $\|T\|^2 = \|T^*T\|$ and of Lemma 5.10 will be played by the following Lemma.

LEMMA 5.12. Let $T \in [\mathscr{X}]$ and $\varepsilon > 0$. For each $y \in T\mathscr{B}$, choose $f_y \in \mathscr{X}^*$ such that

$$\left| f_y(y) - \|y\|^2 \right| \leqq \varepsilon.$$

(The existence of such elements f_y is guaranteed by the Hahn-Banach theorem.) Let $\mathscr{F} = \{f_y : y \in T\mathscr{B}\} \subset \mathscr{X}^*$. Then

$$\|T\|^2 \leqq \sup_{f \in \mathscr{F}} \|T^*f\| + \varepsilon.$$

Proof. Let $x \in \mathscr{B}$ and $y = Tx$. Then

$$f_y(y) = f_y(Tx) = (T^*f_y)x,$$

$$|f_y(y)| \leqq \|T^*f_y\| \leqq \sup_{f \in \mathscr{F}} \|T^*f\|,$$

$$\|Tx\|^2 = \|y\|^2 \leqq |f_y(y)| + \varepsilon \leqq \sup_{f \in \mathscr{F}} \|T^*f\| + \varepsilon.$$

The assertion follows.

With this preparation we are ready to establish Theorem 5.8 in general.

Proof 2. $\mathscr{X}$ a Banach space. Fix $\varepsilon > 0$ and suppose that $\hat{\mathscr{X}}$ is an infinite subset of $\mathscr{X}$. By hypothesis (a) of the theorem, the set

$$\mathscr{S} = \{Kx - Lx : K, L \in \mathscr{K}, \quad x \in \mathscr{B}\}$$

is totally bounded. Define $\mathscr{F}$ as in Lemma 5.11. By hypothesis (b) and Lemma 5.9 there is an infinite set $\mathscr{X}_\varepsilon \subset \hat{\mathscr{X}}$ such that

$$\|(K - L)^*f\| < \varepsilon \quad \text{for} \quad K, L \in \mathscr{X}_\varepsilon, \quad f \in \mathscr{F}.$$

Hence, Lemma 5.12 implies

$$\|K - L\|^2 < 2\varepsilon \quad \text{for} \quad K, L \in \mathscr{X}_\varepsilon.$$

By Lemma 5.9, $\mathscr{X}$ is totally bounded and Theorem 5.8 is proved.

5.7 CONCLUDING REMARKS

Let us recapitulate. Our primary goal was Theorem 5.5: a set $\mathscr{K}$ of compact operators in $[\mathscr{X}]$ is totally bounded iff both $\mathscr{K}$ and $\mathscr{K}^*$ are collectively compact. This follows from Proposition 5.3 and Theorem 5.8.

In Theorem 5.8 the roles of $\mathscr{K}$ and $\mathscr{K}^*$ can be interchanged. For proofs of the following theorems and related results see Palmer [65].

THEOREM 5.13. Let $\mathscr{K} \subset [\mathscr{X}]$. Assume

$\mathscr{K}x$ is totally bounded for each $x \in \mathscr{X}$,

$\mathscr{K}^*$ is collectively compact.

Then $\mathscr{K}$ is totally bounded.

THEOREM 5.14. Let $\mathscr{K} \subset [\mathscr{X}]$. Then $\mathscr{K}^*$ is collectively compact iff $\mathscr{K}$ is bounded and, for each $\varepsilon > 0$, there is a subspace $\mathscr{X}_\varepsilon$ of finite codimension in $\mathscr{X}$ such that

$$\|K_\varepsilon\| \leq \varepsilon \quad \text{for} \quad K \in \mathscr{K},$$

where K_ε is the restriction of K to $\mathscr{X}_\varepsilon$.

THEOREM 5.15. Let $\mathscr{K}$ be a set of compact operators in $[\mathscr{X}]$. Then $\mathscr{K}$ is totally bounded iff $\mathscr{K}x$ is totally bounded for each $x \in \mathscr{X}$ and, for each $\varepsilon > 0$, there is a subspace $\mathscr{X}_\varepsilon$ of finite codimension in $\mathscr{X}$ such that

$$\|K_\varepsilon\| \leq \varepsilon \quad \text{for} \quad K \in \mathscr{K},$$

where K_ε is the restriction of K to $\mathscr{X}_\varepsilon$.

chapter 6

APPROXIMATE SOLUTIONS OF NONLINEAR EQUATIONS

6.1 INTRODUCTION

The approximation theory for linear equations developed in previous chapters will be extended to certain nonlinear equations

$$Px = 0, \qquad P_n x = 0, \qquad n = 1, 2, \ldots,$$

in a Banach space $\mathscr{X}$. Although the analysis is more complicated, the results parallel those for the linear case: relations between the solvability of the equations $Px = 0$ and $P_n x = 0$, convergence of approximate solutions, and error bounds.

Much of the theoretical discussion is motivated by the following important example. Consider the *Urysohn* integral equation in $\mathscr{C} = \mathscr{C}[0,1]$,

$$x(s) - \int_0^1 k(s,t,x(t))\, dt = z(s), \qquad 0 \leq s \leq 1,$$

where $k(s,t,x)$ is continuous in s and t, and is twice continuously differentiable in x. Suppose a prescribed quadrature formula satisfies

$$\sum_{j=1}^n w_{nj} x(t_{nj}) \to \int_0^1 x(t)\, dt, \qquad x \in \mathscr{C}.$$

Numerical integration approximations of the Urysohn equation are given by

$$x(s) - \sum_{j=1}^{n} w_{nj} k(s, t_{nj}, x(t_{nj})) = z(s), \qquad 0 \leq s \leq 1.$$

For each n this equation is essentially equivalent to the nonlinear algebraic system

$$x(t_{ni}) - \sum_{j=1}^{n} w_{nj} k(t_{ni}, t_{nj}, x(t_{nj})) = z(t_{ni}), \qquad i = 1, \ldots, n.$$

We shall put the foregoing equations in a formal setting. Let K be the Urysohn integral operator on $\mathscr{C}$,

$$(Kx)(s) = \int_{0}^{1} k(s, t, x(t)) \, dt, \qquad 0 \leq s \leq 1.$$

Define operators K_n, $n = 1, 2, \ldots$, on $\mathscr{C}$ by

$$(K_n x)(s) = \sum_{j=1}^{n} w_{nj} k(s, t_{nj}, x(t_{nj})), \qquad 0 \leq s \leq 1.$$

Define operators $\hat{K}_n$, $n = 1, 2, \ldots$, on l_n^{∞} by

$$(\hat{K}_n \hat{x})_i = \sum_{j=1}^{n} w_{nj} k(t_{ni}, t_{nj}, \hat{x}_j), \qquad i = 1, \ldots, n.$$

Then the Urysohn integral equation and the corresponding numerical integration approximations are expressed in operator form as

$$Px \equiv (I - K)x - z = 0,$$
$$P_n x \equiv (I - K_n)x - z = 0,$$
$$\hat{P}_n \hat{x} \equiv (\hat{I} - \hat{K}_n)\hat{x} - \hat{z} = 0,$$

with $\hat{x}_j = x(t_{nj})$ and $\hat{z}_j = z(t_{nj})$ in the last equation.

The foregoing example will be referred to from time to time. We return now to the general theory. As might be expected, compactness hypotheses will play key roles in the analysis. The concept of a collectively compact set or sequence is extended to nonlinear (i.e., not necessarily linear) operators as follows.

DEFINITION. A set $\mathscr{K}$ of nonlinear operators on $\mathscr{X}$ is *collectively compact* if for each bounded set $\mathscr{S} \subset \mathscr{X}$ the set $\mathscr{K}\mathscr{S}$ is relatively compact. A sequence of nonlinear operators is collectively compact whenever the corresponding set is collectively compact.

This generalizes the usual definition of a single compact nonlinear operator. The Urysohn integral operator K introduced above is compact. The corresponding sequence $\{K_n\}$ of numerical integration approximations is collectively compact.

As in the integral equation example, the replacement of P by P_n is intended to yield a more tractable problem of the same general type, perhaps finite dimensional. Such discretizing approximations are often necessary as well as desirable when numerical computations are performed. Since ordinarily P_n is genuinely nonlinear when P is, the equations $P_n x = 0$ retain some of the difficulties of the original equation $Px = 0$.

An effective additional tool is Newton's method of successive approximations (described in §6.3) which replaces a nonlinear equation by a sequence of linear equations. Theoretically, Newton's method can be applied directly to $Px = 0$. In practice, it is usually applied to a simpler related equation, such as $P_n x = 0$ or $\hat{P}_n \hat{x} = 0$ in the case of the Urysohn integral equation.

In the general situation, then, the replacement of P by P_n abstracts the notion of discretization, while Newton's method is a linearization procedure. The analysis in this chapter takes account of both "discretization" and "linearization" effects. Most of the results are adapted from [60, 61] by Moore. Related contributions and applications have been made by Anselone and Moore [14], Baluev [24], Bryan [27], Davis [33], Krasnosel'skii and Rutickii [52], and Mysovskih [57, 58].

In a quite different direction from the theme of this chapter, Daniel [32] has investigated collectively compact sets of gradient mappings. His results pertain to the approximate solution of variational problems.

6.2 FRÉCHET DIFFERENTIATION

Newton's method involves differentiability conditions on the operators. A brief resumé of Fréchet differentiation is given in this section. Recommended background references are Dieudonné [37], Kantorovich and Akilov [47], and Rall [68]. In what follows, $\mathcal{X}$ and $\mathcal{Y}$ are Banach spaces, $[\mathcal{X},\mathcal{Y}]$ is the space of bounded linear operators on $\mathcal{X}$ into $\mathcal{Y}$,

$$\mathcal{B}(x,r) = \{x_1 \in \mathcal{X} : \|x_1 - x\| \leq r\}, \qquad x \in \mathcal{X}, \qquad r > 0,$$

and $\mathcal{B} = \mathcal{B}(0,1)$ as before.

Let K be an operator with domain $\mathcal{D}(K) \subset \mathcal{X}$ and range $\mathcal{R}(K) \subset \mathcal{Y}$. Let $x \in \mathcal{D}(K)$ and assume that $\mathcal{B}(x,r) \subset \mathcal{D}(K)$ for some $r > 0$ (the latter condition can be weakened). Then K is *Fréchet differentiable at x* iff there is a bounded linear operator $K'(x) \in [\mathcal{X},\mathcal{Y}]$ such that

$$\lim_{\|u\| \to 0} \frac{\|K(x + u) - Kx - K'(x)u\|}{\|u\|} = 0,$$

in which case $K'(x)$ is the (unique) *Fréchet derivative of K at x*. An equivalent definition of $K'(x)$ is provided by

$$\lim_{\delta \to 0} \left\| \frac{K(x + \delta u) - Kx}{\delta} - K'(x)u \right\| = 0 \quad \text{uniformly for} \quad \|u\| = 1.$$

Without the uniformity of the convergence this would define the Gateaux derivative of K at x. Since all derivatives appearing below are in the sense of Fréchet, the modifier Fréchet will be omitted for the sake of brevity.

Whenever $K'(x)$ exists for at least one $x \in \mathcal{X}$, the *derivative* of K is defined as the operator K' with domain

$$\mathcal{D}(K') = \{x \in \mathcal{X} : \exists K'(x)\}$$

and with values $K'(x)$. Thus, $\mathcal{D}(K') \subset \mathcal{X}$ and $\mathcal{R}(K') \subset [\mathcal{X},\mathcal{Y}]$.

Whenever $(K')'$ exists, the *second derivative* of K is defined as $K'' = (K')'$. Then $\mathcal{D}(K'') \subset \mathcal{X}$ and $\mathcal{R}(K'') \subset [\mathcal{X},[\mathcal{X},\mathcal{Y}]]$. Hence, for $x \in \mathcal{D}(K'')$ and $u,v \in \mathcal{X}$,

$$K''(x) \in [\mathcal{X},[\mathcal{X},\mathcal{Y}]],$$
$$K''(x)u \in [\mathcal{X},\mathcal{Y}],$$
$$K''(x)uv \in \mathcal{Y},$$
$$\|K''(x)\| = \sup_{u,v \in \mathcal{B}} \|K''(x)uv\|.$$

It is not difficult to show that

$$K''(x)uv = K''(x)vu.$$

Often $K''(x)$ is regarded as a symmetric bilinear operator on $\mathcal{X} \times \mathcal{X}$ into $\mathcal{Y}$.

Higher derivatives are defined by induction: $K^{(n+1)} = (K^{(n)})'$. Only the first and second derivatives will be needed below.

Some of the more important properties of derivatives are the following. If K is differentiable at x, then K is continuous at x. If K is constant then $K' = O$. If $K \in [\mathcal{X},\mathcal{Y}]$, then $K' = K$ and $K'' = O$. The map $K \mapsto K'$ is linear in the sense that

$$(c_1 K_1 + c_2 K_2)'(x) = c_1 K_1'(x) + c_2 K_2'(x)$$

when K_1 and K_2 are differentiable at x. The derivative of a composition K_2K_1 is given by the chain rule,

$$(K_2K_1)'(x) = K_2'(K_1x)K_1'(x),$$

when K_1 is differentiable at x and K_2 is differentiable at K_1x. In one of its various forms, the first mean value theorem states that if K is differentiable on $\mathscr{S} = \{x + \delta u: 0 \leq \delta \leq 1\}$ then

$$\|K(x + u) - Kx\| \leq \|u\| \sup_{y \in \mathscr{S}} \|K'(y)\|.$$

Forms of the second mean value theorem are

$$\|K(x + u) - Kx - K'(x)u\| \leq \tfrac{1}{2} \|u\| \sup_{y \in \mathscr{S}} \|K'(y) - K'(x)\|,$$

$$\|K(x + u) - Kx - K'(x)u\| \leq \tfrac{1}{2} \|u\|^2 \sup_{y \in \mathscr{S}} \|K''(y)\|,$$

when the derivatives exist.

As an example, let K be the Urysohn integral operator on $\mathscr{C}$ introduced in §6.1. Then $K'(x)$ is the linear integral operator on $\mathscr{C}$ defined by

$$[K'(x)u](s) = \int_0^1 \frac{\partial k}{\partial x}(s,t,x(t))u(t)\,dt$$

and $K''(x)$ is given by

$$[K''(x)uv](s) = \int_0^1 \frac{\partial^2 k}{\partial x^2}(s,t,x(t))u(t)v(t)\,dt.$$

The numerical integral operators K_n have the first and second derivatives determined by

$$[K_n'(x)u](s) = \sum_{j=1}^n w_{nj} \frac{\partial k}{\partial x}(s,t_{nj},x(t_{nj}))u(t_{nj}),$$

$$[K_n''(x)uv](s) = \sum_{j=1}^n w_{nj} \frac{\partial^2 k}{\partial x^2}(s,t_{nj},x(t_{nj}))u(t_{nj})v(t_{nj}).$$

Note that $K_n'(x)$ and $K_n''(x)$ are numerical integration approximations of $K'(x)$ and $K''(x)$. Thus, differentiation and discretization can be performed in either order.

Now suppose that K is a twice differentiable operator on a Banach space $\mathscr{X}$. Define

$$Px \equiv (I - K)x - z,$$

where $z \in \mathscr{X}$. Then

$$P'(x) = I - K'(x), \qquad P''(x) = -K''(x).$$

The Urysohn integral equation and the corresponding numerical integration approximations are special cases of

$$Px \equiv (I - K)x - z = 0.$$

We shall make a detailed study of such equations after some further preparatory material.

6.3 NEWTON'S METHOD

Let P be any differentiable operator on $\mathscr{X}$ and consider the equation

$$Px = 0.$$

For each $x_0 \in \mathscr{X}$,

$$Px = Px_0 + P'(x_0)(x - x_0) + R(x),$$

where $R(x)/\|x - x_0\| \to 0$ as $\|x - x_0\| \to 0$. This suggests that if x_0 is near a solution of $Px = 0$, then a better approximation x_1 might be obtained from the linear equation

$$Px_0 + P'(x_0)(x_1 - x_0) = 0,$$

which has the unique solution

$$x_1 = x_0 - P'(x_0)^{-1}Px_0$$

when $P'(x_0)^{-1} \in [\mathscr{X}]$.

Repetition of the foregoing procedure yields the iteration formula

$$x_{m+1} = x_m - P'(x_m)^{-1}Px_m, \qquad m = 0, 1, 2, \ldots,$$

which is the basis of *Newton's method* for the successive approximate solution of $Px = 0$. Although the method does depend materially on the existence of $P'(x_m)^{-1} \in [\mathscr{X}]$ for $m \geq 0$, it is not necessary to evaluate the operator inverses. Instead, it suffices to solve the equation for each x_{m+1},

$$Px_m + P'(x_m)(x_{m+1} - x_m) = 0, \qquad m = 0, 1, 2, \ldots.$$

If $Px = 0$ is the Urysohn integral equation, then these are Fredholm integral equations of the second kind.

The next theorem gives sufficient conditions for the success of Newton's method. First, some convenient notation:

$$\omega(h) = \frac{1 - \sqrt{1 - 2h}}{h} \quad \text{for} \quad 0 < h \leq \tfrac{1}{2}, \qquad \omega(0) = 1,$$

$$\omega^{+}(h) = \frac{1 + \sqrt{1 - 2h}}{h} \quad \text{for} \quad 0 < h \leq \tfrac{1}{2}.$$

Hypotheses stated in terms of derivatives or inverses include the existence of those operators.

THEOREM 6.1. Let P be an operator on $\mathcal{X}$ and $x_0 \in \mathcal{X}$. Assume $P'(x_0)^{-1} \in [\mathcal{X}]$ and

$$\|P'(x_0)^{-1}\| \leq \xi, \qquad \|P'(x_0)^{-1}Px_0\| \leq \eta, \qquad \|P''(x)\| \leq \zeta \quad \text{on} \quad \mathcal{B}(x_0, r),$$

$$h = \xi\eta\zeta \leq \tfrac{1}{2}, \qquad r_0 = \eta\omega(h) \leq r.$$

Then there is a unique $x^* \in \mathcal{B}(x_0, r_0)$ such that

$$Px^* = 0.$$

If $h < \tfrac{1}{2}$ and $r \geq r_1 = \eta\omega^{+}(h)$, then x^* is the unique solution of $Px = 0$ with $\|x - x_0\| < r_1$. The Newton formula,

$$x_{m+1} = x_m - P'(x_m)^{-1}Px_m, \qquad m = 0, 1, 2, \ldots,$$

defines a sequence in $\mathcal{B}(x_0, r_0)$ such that $P'(x_m)^{-1} \in [\mathcal{X}]$ and

$$\|x_m - x^*\| \leq \frac{1}{2^{m-1}}(2h)^{2^m - 1}\eta \to 0 \quad \text{as} \quad m \to \infty.$$

Except for minor changes, this important theorem is due to Kantorovich [46]. Well organized proofs are given by Kantorovich and Akilov [47], Antosiewicz and Rheinboldt [18], and Rall [68].

The hypothesis $\|P'(x_0)^{-1}Px_0\| \leq \eta$ in Theorem 6.1 is expressed in terms of Newton iterates by $\|x_1 - x_0\| \leq \eta$. If $\|Px_0\| \leq \beta$, then the theorem remains valid with $\eta = \xi\beta$ and $h = \xi^2\beta\zeta$. Since $\omega(h) \leq 2$ for $0 \leq h \leq \tfrac{1}{2}$, the inequality $r_0 \leq r$ is satisfied whenever $r \geq 2\eta$.

The *modified Newton's method*, based on the iteration formula

$$x_{m+1} = x_m - P'(x_0)^{-1} Px_m, \qquad m = 0, 1, 2, \ldots,$$

is easier to use in practice, but converges more slowly for most cases.

COROLLARY 6.2. Let $h < \frac{1}{2}$ and $r_0 \leq r$ in Theorem 6.1. Then there exists $P'(x^*)^{-1} \in [\mathscr{X}]$ and

$$\|P'(x^*)^{-1}\| \leq \frac{\xi}{\sqrt{1 - 2h}}, \qquad \|P'(x_m)^{-1}\| \leq \frac{\xi}{\sqrt{1 - 2h}}$$

for all m.

Proof. Application of the first mean value theorem to P' yields, for $x \in \mathscr{B}(x_0, r_0)$,

$$\|P'(x_0)^{-1}\| \, \|P'(x) - P'(x_0)\| \leq \xi \zeta r_0 = h\omega(h) = 1 - \sqrt{1 - 2h} < 1.$$

The desired results follow from Proposition 1.3.

It is sometimes a major task to obtain an approximate solution x_0 of $Px = 0$ which satisfies the hypotheses of Theorem 6.1. One way to attempt this is to apply Newton's method without reference to the theorem until an apparently very good approximate solution is obtained, which is then relabeled x_0. Alternatively, Newton's method may be applied to a more tractable related equation. For example, let $Px = 0$ be the Urysohn integral equation of §6.1. Then Newton's method could be applied either to $P_n x = 0$ or to $\hat{P}_n \hat{x} = 0$ with n fixed reasonably large. In the latter case, if a good approximate solution of $\hat{P}_n \hat{x} = 0$ has been obtained, then a logical candidate for x_0 is

$$x_0(s) = \sum_{j=1}^{n} w_{nj} \, k(s, t_{nj}, \hat{x}_j) + z(s).$$

Another means for obtaining a suitable approximate solution x_0 of $Px = 0$ is "practical homotopy." Suppose that P_λ is an operator on $\mathscr{X}$ for $0 \leq \lambda \leq 1$ which depends continuously on λ, $P_0 x = 0$ is solvable, and $P_1 = P$. For example,

$$P_\lambda = (1 - \lambda)I + \lambda P.$$

Let $0 = \lambda_0 < \lambda_1 < \cdots < \lambda_m = 1$. Then it may be possible to solve $P_{\lambda_i} x = 0, i = 1, \ldots, m$, by Newton's method, using the solution of $P_{\lambda_{i-1}} x = 0$ as the initial guess. This scheme is especially appropriate if there is a parameter already present, as is the case in many applied problems. If successful, it gives approximate values of a *function* $\{x_\lambda : 0 \leq \lambda \leq 1\}$.

There is serious difficulty with the foregoing application of Newton's method if $P'_\lambda(x_\lambda)$ is singular for some $\lambda \in [0, 1]$. In that event, however, a change of parameter $\lambda = \lambda(x, \mu)$ may transform the problem into one for which a singular derivative does not occur. Such a transformation was used by Anselone and Moore [14] in connection with a problem from nonlinear elasticity theory: the buckling of a thin spherical cap. The change of parameter technique was put on a solid theoretical foundation and was extended to multidimensional parameters by Davis [33].

Suppose that, by whatever means, an element x_0 has been found for which the hypotheses of Theorem 6.1 are satisfied. Then $Px^* = 0$ for some x^* such that $\|x^* - x_0\| \leq r_0$. If r_0 is small enough, then x_0 itself is an acceptable approximate solution of $Px = 0$, and no (further) iterations are needed. For this reason, the existence and uniqueness parts of the theorem are ordinarily invoked after all other calculations have been performed in the effort to solve $Px = 0$.

No matter how x_0 is determined, $P'(x_0)^{-1}$ is needed in Theorem 6.1. Usually it must be approximated. The theory developed in previous chapters is available for this purpose. Thus, suppose there exist operators P_n, $n = 1, 2, \ldots$, on $\mathscr{X}$ such that for each $x \in \mathscr{X}$,

$$P'_n(x) \to P'(x),$$

$$\{P'_n(x) - P'(x)\} \text{ is collectively compact.}$$

Then ξ and η for use in Theorem 6.1 can be obtained from Theorem 4.6 with $\lambda = 0$, provided that $P'_n(x_0)^{-1}$ can be dealt with. Similarly, if

$$Px \equiv (I - K)x - z, \qquad P_n x \equiv (I - K_n)x - z,$$

and, for each $x \in \mathscr{X}$,

$$K'_n(x) \to K'(x),$$

$$\{K'_n(x)\} \text{ is collectively compact,}$$

then ξ and η can be obtained from Theorem 1.12.

Bryan [27] combined the inequalities in Theorem 1.12 and 6.1 to obtain a variant of the Kantorovich theorem which deals directly with the present situation. He also presented an example based on the integral equation

$$x(s) - \int_0^1 e^{st - x^2(t)} \, dt = 1.$$

The following two sections are devoted to a study of conditions on an operator P of the form $Px \equiv (I - K)x - z$, on a perturbation $\tilde{P}$ of P, and on an approximate solution of $\tilde{P}x = 0$, which guarantee the existence of a nearby exact solution of $Px = 0$. In §6.6, $\tilde{P}$ is replaced by a sequence of operators P_n which converge to P in a certain sense. The material is adapted with alterations from Moore [61].

6.4 VARIATIONS OF THE KANTOROVICH THEOREM

Many variations of the Kantorovich theorem have appeared in the literature. Moore [59] gives a recent summary. The successive variations given below culminate in one designed specifically for our purposes.

THEOREM 6.3. Let P be an operator on $\mathscr{X}$ and $x_0 \in \mathscr{X}$. Assume $P'(x_0)^{-1} \in [\mathscr{X}]$ and

$$\|P'(x_0)^{-1}Px_0\| \leq \eta,$$

$$\|P'(x_0)^{-1}[P'(x) - P'(y)]\| \leq \kappa \|x - y\| \quad \text{on} \quad \mathscr{B}(x_0, r),$$

$$h = \eta \kappa \leq \tfrac{1}{2}, \qquad r_0 = \eta \omega(h) \leq r.$$

Then there is a unique $x^* \in \mathscr{B}(x_0, r_0)$ such that $Px^* = 0$. The Newton iterates x_m are defined, $P'(x_m)^{-1} \in [\mathscr{X}]$ for all m, and $\|x_m - x^*\| \to 0$.

This is proved in very much the same way as Theorem 6.1 (cf. Davis [33], Dennis [35], Kantorovich and Akilov [47]). The other conclusions of Theorems 6.1 and 6.2 carry over with appropriate notational changes to Theorem 6.3 and to others which follow.

THEOREM 6.4. Let P be an operator on $\mathscr{X}$, $G \in [\mathscr{X}]$, and $x_0 \in \mathscr{X}$. Assume $I - P'(x_0)$ is compact and

$$\|I - GP'(x_0)\| \leq \tau < 1, \qquad\qquad \|GPx_0\| \leq \eta,$$
$$\|G[P'(x) - P'(y)]\| \leq \kappa \|x - y\| \quad \text{on} \quad \mathscr{B}(x_0, r),$$
$$h = \frac{\eta\kappa}{(1 - \tau)^2} \leq \frac{1}{2}, \qquad\qquad r_0 = \frac{\eta\omega(h)}{1 - \tau} \leq r.$$

Then there is a unique $x^* \in \mathscr{B}(x_0, r_0)$ such that $Px^* = 0$. The Newton iterates x_m are defined, $P'(x_m)^{-1} \in [\mathscr{X}]$ for all m, and $\|x_m - x^*\| \to 0$.

Proof. From the first inequality of the theorem and Proposition 1.1,

$$[GP'(x_0)]^{-1} \in [\mathscr{X}], \qquad \|[GP'(x_0)]^{-1}\| \leq \frac{1}{1 - \tau}.$$

Therefore $P'(x_0)^{-1}$ exists. Since $I - P'(x_0)$ is compact, the Fredholm alternative implies that $P'(x_0)^{-1} \in [\mathscr{X}]$. Since $P'(x_0)^{-1} = [GP'(x_0)]^{-1}G$, the hypotheses of Theorem 6.3 are satisfied with η and κ replaced by $\eta/(1 - \tau)$ and $\kappa/(1 - \tau)$.

Since the compactness of $I - P'(x_0)$ was used only for the Fredholm alternative, that hypothesis can be weakened.

In the next theorem, P is of the form $Px \equiv (I - K)x - z$ and M is an approximation to $K'(x_0)$ at some $x_0 \in \mathscr{X}$. Since $P'(x) = I - K'(x)$, hypotheses involving derivatives can be put on P or K. We have chosen the latter.

THEOREM 6.5. Let K be an operator on $\mathscr{X}$, $z \in \mathscr{X}$, and

$$Px \equiv (I - K)x - z.$$

Let $M \in [\mathscr{X}]$ and $x_0 \in \mathscr{X}$. Assume $K'(x_0)$ is compact and

$$(I - M)^{-1} \in [\mathscr{X}], \qquad \|(I - M)^{-1}\| \leq \beta, \qquad \|Px_0\| \leq d,$$
$$\|[M - K'(x_0)]Px_0\| \leq d_0, \qquad \|[M - K'(x_0)]K'(x_0)\| \leq d_1 < \frac{1}{\beta},$$
$$\|K'(x) - K'(y)\| \leq \kappa \|x - y\| \qquad\qquad \text{on} \quad \mathscr{B}(x_0, r),$$
$$\|[M - K'(x_0)][K'(x) - K'(y)]\| \leq d_2 \|x - y\| \quad \text{on} \quad \mathscr{B}(x_0, r),$$
$$h = \frac{\beta^2(d + d_0)(\kappa + d_2)}{(1 - \beta d_1)^2} \leq \frac{1}{2}, \qquad r_0 = \frac{\beta(d + d_0)\omega(h)}{1 - \beta d_1} \leq r.$$

Then there is a unique $x^* \in \mathscr{B}(x_0, r_0)$ such that $Px^* = 0$. The Newton iterates x_m are defined, $P'(x_m)^{-1} \in [\mathscr{X}]$ for all m, and $\|x_m - x^*\| \to 0$.

Proof. Define

$$G = I + (I - M)^{-1}K'(x_0) = (I - M)^{-1}[I - M + K'(x_0)].$$

Elementary manipulations yield

$$GP'(x_0) - I = (I - M)^{-1}[M - K'(x_0)]K'(x_0),$$

$$\|GP'(x_0) - I\| \le \beta d_1 < 1.$$

The hypotheses of Theorem 6.4 are satisfied with τ, η, κ replaced by βd_1, $\beta(d + d_0)$, $\beta(\kappa + d_2)$.

The steps which yield $GP'(x_0) - I$ are much the same as those used in the proof of Theorem 1.10. Thus, in effect, the linear operator approximation theory is now built into the nonlinear theory. Incidentally, the existence of $P'(x_0)^{-1} \in [\mathscr{X}]$ can be deduced either from Theorem 6.4 or from Theorem 1.10. See Moore [61] for further details.

6.5 A PERTURBATION THEOREM

Consider equations on $\mathscr{X}$ of the form

$$Px \equiv (I - K)x - z = 0, \qquad \tilde{P}x \equiv (I - \tilde{K})x - z = 0.$$

For example, $Px = 0$ might be a Urysohn integral equation and $\tilde{P}x = 0$ a corresponding numerical integration approximation (or vice versa). The next theorem implies that if $\tilde{K}$ approximates K well enough and if $\tilde{x}$ is a sufficiently good approximate solution of $\tilde{P}x = 0$, then the equation $Px = 0$ has a solution near $\tilde{x}$. In spite of the apparent complexity of the theorem the hypotheses should not be too difficult to check in practice (cf. §6.6).

THEOREM 6.6. Let K and $\tilde{K}$ be operators on $\mathscr{X}$, $z \in \mathscr{X}$, and

$$Px \equiv (I - K)x - z, \qquad \tilde{P}x \equiv (I - \tilde{K})x - z.$$

Let $R \geq r > 0$, $\mathscr{V} \subset \mathscr{X}$, and $\mathscr{V}_r = \bigcup_{x \in \mathscr{V}} \mathscr{B}(x,r)$. Assume

$$\tilde{K}\mathscr{B}(0,R) + z \subset \mathscr{V}, \qquad K'(x) \quad \text{compact for} \quad x \in \mathscr{V}_r,$$

$$\|K'(x) - K'(y)\| \leq \kappa \|x - y\|, \qquad \|\tilde{K}'(x) - \tilde{K}'(y)\| \leq \kappa \|x - y\|$$

$$\text{for} \quad x,y \in \mathscr{V}_r,$$

$$\|\tilde{K}x - Kx\| \leq d, \qquad \|[\tilde{K}'(x) - K'(x)]Px\| \leq d_0,$$

$$\|[\tilde{K}'(x) - K'(x)]K'(x)\| \leq d_1 \quad \text{for} \quad x \in \mathscr{V},$$

$$\|[\tilde{K}'(x) - K'(x)][K'(y) - K'(\bar{y})]\| \leq d_2 \|y - \bar{y}\|$$

$$\text{for} \quad x \in \mathscr{V}, \; y,\bar{y} \in \mathscr{V}_r.$$

Fix $x \in \mathscr{X}$ and assume

$$\|\tilde{x}\| \leq R - r, \qquad \|\tilde{P}\tilde{x}\| \leq \rho, \qquad \tilde{P}'(\tilde{x})^{-1} \in [\mathscr{X}], \qquad \|\tilde{P}'(\tilde{x})^{-1}\| \leq \beta,$$

$$\tilde{h} = \beta^2 \kappa \rho, \qquad 2\tilde{h} + \beta^2 d_1^2 < 1, \qquad \tilde{r}_0 = \beta \rho \omega(\tilde{h}), \qquad \tilde{r}_0 + \rho \leq r,$$

$$h = \frac{\beta^2(d + d_0)(\kappa + d_2)}{(\sqrt{1 - 2\tilde{h}} - \beta d_1)^2} \leq \frac{1}{2}, \qquad r_0 = \frac{\beta(d + d_0)\omega(h)}{\sqrt{1 - 2\tilde{h}} - \beta d_1} \leq r.$$

Then there is a unique $\tilde{x}^* \in \mathscr{B}(\tilde{x},\tilde{r}_0)$ such that $\tilde{P}\tilde{x}^* = 0$ and there is a unique $x^* \in \mathscr{B}(\tilde{x}^*,r_0)$ such that $Px^* = 0$.

Proof. From $\|\tilde{x}\| \leq R$ and $\tilde{K}\mathscr{B}(0, R) + z \subset \mathscr{V}$ it follows that $\tilde{K}\tilde{x} + z \in \mathscr{V}$. Note that $\rho \leq r$. If $\|x - \tilde{x}\| \leq r - \rho$, then

$$\|x - (\tilde{K}\tilde{x} + z)\| \leq$$

$$\|x - \tilde{x}\| + \|\tilde{x} - \tilde{K}\tilde{x} - z\| = \|x - \tilde{x}\| + \|\tilde{P}\tilde{x}\| \leq r$$

and $x \in \mathscr{V}_r$. Thus, $\mathscr{B}(x,r - \rho) \subset \mathscr{V}_r$. Now

$$\|\tilde{P}'(\tilde{x})^{-1}\tilde{P}\tilde{x}\| \leq \beta\rho,$$

$$\|\tilde{P}'(\tilde{x})^{-1}[\tilde{P}'(x) - \tilde{P}'(y)]\| \leq \beta\kappa \|x - y\| \quad \text{on} \quad \mathscr{B}(\tilde{x},r - \rho),$$

$$\tilde{h} = \beta^2 \kappa \rho < \tfrac{1}{2}, \qquad \tilde{r}_0 = \beta\rho\omega(\tilde{h}) \leq r - \rho.$$

By Theorem 6.3 there is a unique $\tilde{x}^* \in \mathscr{B}(\tilde{x},\tilde{r}_0)$ such that $\tilde{P}\tilde{x}^* = 0$. Since $\tilde{h} < \tfrac{1}{2}$ (cf. Corollary 6.2),

$$\tilde{P}'(\tilde{x}^*)^{-1} \in [\mathscr{X}], \qquad \|\tilde{P}'(\tilde{x}^*)^{-1}\| \leq \frac{\beta}{\sqrt{1 - 2\tilde{h}}}.$$

Since $\|\tilde{x}^*\| \leq \|\tilde{x}\| + \tilde{r}_0 \leq R - \rho$ and $\tilde{x}^* = \tilde{K}\tilde{x}^* + z \in \mathscr{V}$, we have $\mathscr{B}(\tilde{x}^*,r) \subset \mathscr{V}_r$. The hypotheses of Theorem 6.5 are satisfied with x_0, M, β replaced by $\tilde{x}^*$, $\tilde{K}'(\tilde{x}^*)$, $\beta/\sqrt{1 - 2\tilde{h}}$. The verifications are straightforward except for

$$\|P\tilde{x}^*\| \leq \|P\tilde{x}^* - \tilde{P}\tilde{x}^*\| + \|\tilde{P}\tilde{x}^*\| = \|K\tilde{x}^* - \tilde{K}\tilde{x}^*\| \leq d,$$

where we have used $\tilde{P}\tilde{x}^* = 0$ and $\tilde{x}^* \in \mathscr{V}$.

The theorem can be augmented in a number of ways. First,

$$\|x^* - \tilde{x}\| \leq r_0 + \tilde{r}_0 \leq 2r - \rho, \qquad \|x^*\| \leq R + r - \rho.$$

If either $r \leq \rho$ or $\|\tilde{x}\| \leq R - 2r$ then $\|x^*\| \leq R$, which is necessary if R is an *a priori* bound on admissible solutions of $Px = 0$. Both $\tilde{x}^*$ and x^* are limits of Newton sequences with error bounds inherited from Theorem 6.1. Since $\tilde{h} < \frac{1}{2}$, there is a larger uniqueness region for $\tilde{x}^*$. If $h < \frac{1}{2}$, then the same can be said for x^* and, in addition, $P'(x^*)^{-1} \in [\mathscr{X}]$. It follows from the mean value theorem that the inequalities in Theorem 6.6 involving κ and d_2 are satisfied if

$$\|K''(x)\| \leq \kappa, \qquad \|K''(x)\| \leq \kappa \quad \text{for} \quad x \in \mathscr{V}_r,$$

$$\|[\tilde{K}'(x) - K'(x)]K''(y)\| \leq d_2 \quad \text{for} \quad x \in \mathscr{V}, \qquad y \in \mathscr{V}_r.$$

If $\|K'(x)\| \leq c$ and $\|\tilde{K}'(x)\| \leq c$ for $x \in \mathscr{V}_r$, then $d_2 = 2c\kappa$ suffices.

Fix R, r, $\mathscr{V}$, κ and β in Theorem 6.6. Then the inequalities on $\tilde{h}$, $\tilde{r}_0$, h and r are satisfied, and the conclusions of the theorem follow if d, d_0, d_1, d_2 and ρ are small enough. The parameters d, d_0, d_1 and d_2 measure the deviation of $\tilde{K}$ from K, while the size of ρ indicates how good an approximate solution of $\tilde{P}x = 0$ is $\tilde{x}$. Two special cases are of particular interest. If $\tilde{K} = K$ let $d = d_0 = d_1 = d_2 = 0$; then $h = 0$, $r_0 = 0$, and $x^* = \tilde{x}^*$. If $\tilde{P}\tilde{x} = 0$, let $\rho = 0$; then $\tilde{h} = 0$, $\tilde{r}_0 = 0$ and $\tilde{x}^* = \tilde{x}$.

6.6 CONVERGENCE OF APPROXIMATE SOLUTIONS

Let K and K_n, $n = 1, 2, \ldots$, be operators on $\mathscr{X}$, $z \in \mathscr{X}$, and

$$Px \equiv (I - K)x - z; \qquad P_n x \equiv (I - K_n)x - z.$$

Throughout this section the operators K and K_n will satisfy the following hypotheses for any $x \in \mathscr{X}$ and any bounded set $\mathscr{S} \subset \mathscr{X}$:

$$K_n \to K, \qquad K'_n(x) \to K'(x),$$

$\{K_n: n \geq 1\}$ and $\{K'_n(x): n \geq 1\}$ is collectively compact,

$\{K''(x)uv: x,u,v \in \mathscr{S}\}$ and $\{K''_n(x)uv: x,u,v \in \mathscr{S}, n \geq 1\}$
$$\text{are relatively compact.}$$

Examples are furnished by the Urysohn operator and the corresponding numerical integration approximations.

Several fairly direct consequences of the foregoing hypotheses will be stated without proof. For any $x,u \in \mathscr{X}$ and any bounded set $\mathscr{S} \subset \mathscr{X}$,

K and $K'(x)$ are compact,

$\{K'_n(x): n \geq 1\}$ is bounded,

$\{K''(x): x \in \mathscr{S}\}$ and $\{K''_n(x): x \in \mathscr{S}, n \geq 1\}$ are bounded,

K' is continuous, $\{K'_n: n \geq 1\}$ is equicontinuous,

$\{K'_n(y): y \in \mathscr{B}(x,r_x), n \geq 1\}$ is bounded for some $r_x > 0$,

K is continuous, $\{K_n: n \geq 1\}$ is equicontinuous,

the maps $(x,u) \mapsto K'_n(x)u, n = 1, 2, \ldots$, are equicontinuous.

The principal tools in the derivations of these results are the mean value theorem, the Banach-Steinhaus theorem, and the triangle inequality.

Fix $R > 0$ and let

$$\mathscr{V} = \overline{\{K_n x + z: \|x\| \leq R, n \geq 1\}}.$$

Since $K_n \to K$,

$$\mathscr{V} \supset \{Kx + z: \|x\| \leq R\}.$$

Since $\{K_n: n \geq 1\}$ is collectively compact, $\mathscr{V}$ is compact, hence bounded. Choose any r such that $0 < r \leq R$. Define

$$\mathscr{V}_r = \bigcup_{x \in \mathscr{V}} \mathscr{B}(x,r).$$

Since $\mathscr{V}_r$ is bounded, there exists $\kappa < \infty$ such that

$$\|K''(x)\| \leq \kappa, \qquad \|K''_n(x)\| \leq \kappa \quad \text{for} \quad n \geq 1, \qquad x \in \mathscr{V}r.$$

PROPOSITION 6.7. With K and K_n as above,

$$d_n = \sup_{x \in \mathscr{V}} \| K_n x - K x \| \to 0,$$

$$d_{n0} = \sup_{x \in \mathscr{V}} \| [K_n'(x) - K'(x)] P x \| \to 0,$$

$$d_{n0}^* = \sup_{x \in \mathscr{V}} \| [K_n'(x) - K'(x)] P_n x \| \to 0,$$

$$d_{n1} = \sup_{x \in \mathscr{V}} \| [K_n'(x) - K'(x)] K'(x) \| \to 0,$$

$$d_{n1}^* = \sup_{x \in \mathscr{V}} \| [K_n'(x) - K'(x)] K_n'(x) \| \to 0,$$

$$d_{n2} = \sup_{x \in \mathscr{V}, y \in \mathscr{V}_r} \| [K_n'(x) - K'(x)] K''(y) \| \to 0,$$

$$d_{n2}^* = \sup_{x \in \mathscr{V}, y \in \mathscr{V}_r} \| [K_n'(x) - K'(x)] K_n''(y) \| \to 0.$$

Proof. Recall that pointwise convergence of an equicontinuous sequence is uniform on any compact set $\mathscr{U} \subset \mathscr{X}$. Thus, $d_n \to 0$ and

$$\| [K_n'(x) - K'(x)] u \| \to 0 \quad \text{uniformly for} \quad x \in \mathscr{V}, u \in \mathscr{U},$$

from which the other assertions follow.

Fix $\varepsilon > 0$. For some n replace $\tilde{P}, \tilde{K}, d, d_0, d_1, d_2$ by $P_n, K_n, d_n, d_{n0},$ d_{n1}, d_{n2} in Theorem 6.6. If $d_n, d_{n0}, d_{n1}, d_{n2}$ and ρ are sufficiently small, then the theorem yields $x^* \in \mathscr{X}$ such that $P x^* = 0$ and

$$\| x^* - \tilde{x} \| \leq \frac{\beta(d_n + d_{n0}) \omega(h_n)}{1 - \beta d_{n1}} < \varepsilon,$$

where

$$h_n = \frac{\beta^2 (d_n + d_{n0})(\kappa + d_{n2})}{(1 - \beta d_{n1})^2} \leq \frac{1}{2}.$$

These quantities can be calculated in the case of the approximate solution of the Urysohn integral equation, and then $d_n, d_{n0}, d_{n1}, d_{n2}$ are quadrature errors (cf. Chapter 2, §2.5).

Now reverse the roles of P and P_n to obtain the following theorem.

THEOREM 6.8. Assume

$$Px^* = 0, \qquad P'(x^*)^{-1} \in [\mathscr{X}], \qquad \|P'(x^*)^{-1}\| \leq \beta.$$

Then for all n sufficiently large there exist $x_n \in \mathscr{X}$ such that

$$P_n x_n = 0, \qquad \|x_n - x^*\| \to 0.$$

Moreover,

$$\|x_n - x^*\| \leq \frac{\beta(d_n + d_{n0}^*)\omega(h_n^*)}{1 - \beta d_{n1}^*} \to 0,$$

where

$$h_n^* = \frac{\beta^2(d_n + d_{n0}^*)(\kappa + d_{n2}^*)}{(1 - \beta d_{n1}^*)^2} \to 0.$$

Other conclusions can be read off from Theorem 6.6 and the remarks following it. A numerical example based on the Urysohn equation

$$x(s) - \int_0^1 stx^2(t)\, dt = \tfrac{3}{4}s$$

is given by Moore [61].

6.7 EQUICONTINUITY, EQUIDIFFERENTIABILITY, AND COLLECTIVE COMPACTNESS

In the concluding section of the chapter, several properties of operators are related. The results bear directly on the situation investigated in §6.6. The material is from Moore [60].

Various properties of a single operator K on $\mathscr{X}$ generalize more or less directly to uniform properties for a set $\mathscr{K}$ of operators on $\mathscr{X}$. Thus, $\mathscr{K}$ may be equicontinuous or collectively compact. Similarly, $\mathscr{K}$ is equidifferentiable at $x \in \mathscr{X}$ if each $K \in \mathscr{K}$ is differentiable at x and the limit which defines $K'(x)$ is uniform with respect to $K \in \mathscr{K}$.

The mean value theorems yield:

PROPOSITION 6.9. Let $\mathscr{K}$ be a set of operators on $\mathscr{X}$, $x \in \mathscr{X}$, and $r > 0$. Then

$\mathscr{K}$ equidifferentiable at x, $\{K'(x): K \in \mathscr{K}\}$ bounded
$\Rightarrow \mathscr{K}$ equicontinuous at x;

$\{K': K \in \mathscr{K}\}$ equicontinuous at x
$\Rightarrow \mathscr{K}$ equidifferentiable at x;

$\{K''(y): K \in \mathscr{K}, y \in \mathscr{B}(x,r)\}$ bounded
$\Rightarrow \mathscr{K}$ equidifferentiable at x.

According to a theorem of Krasnosel'skii [51], if K is compact and $K'(x)$ exists then $K'(x)$ is compact. This generalizes as follows.

THEOREM 6.10. Let $\mathscr{K}$ be a set of operators on $\mathscr{X}$ and $x \in \mathscr{X}$. Assume

$\mathscr{K}$ collectively compact,
$\mathscr{K}$ equidifferentiable at x.

Then $\{K'(x): K \in \mathscr{K}\}$ is collectively compact.

Proof. Fix $\varepsilon > 0$. Then there exists $\delta > 0$ such that

$$\left\| \frac{K(x + \delta u) - Kx}{\delta} - K'(x)u \right\| < \varepsilon \|u\| \leq \varepsilon$$

for $K \in \mathscr{K}$ and $u \in \mathscr{B}$, the unit ball in $\mathscr{X}$. Therefore, $\{K'(x): K \in \mathscr{K}\}\mathscr{B}$ has the ε-net

$$\left\{ \frac{K(x + \delta u) - Kx}{\delta} : K \in \mathscr{K}, u \in \mathscr{B} \right\}$$

which is totally bounded. It follows that $\{K'(x): K \in \mathscr{K}\}\mathscr{B}$ is totally bounded and that $\{K'(x): K \in \mathscr{K}\}$ is collectively compact.

In view of Proposition 6.9 and Theorem 6.10, the hypothesis that $\{K'_n(x): n \geq 1\}$ is collectively compact in §6.6 is a consequence of the other assumptions. A generalization of Theorem 6.10, proved in the same way, asserts that if $\mathscr{K}$ is collectively compact and is equidifferentiable uniformly on a set $\mathscr{S} \subset \mathscr{X}$, then $\{K'(x): K \in \mathscr{K}, x \in \mathscr{S}\}$ is collectively compact.

THEOREM 6.11. Let K and K_n, $n = 1, 2, \ldots$, be operators on $\mathscr{X}$, $x \in \mathscr{X}$, and $r > 0$. Assume

$$\{K_n : n \geq 1\} \quad \text{or} \quad \{K_n'(x) : n \geq 1\} \text{ collectively compact,}$$
$$\{K_n : n \geq 1\} \quad \text{equidifferentiable at } x,$$
$$K_n \to K \quad \text{on} \quad \mathscr{B}(x, r).$$

Then K is differentiable at x and

$$K_n'(x) \to K'(x).$$

Proof. By Theorem 6.10, $\{K_n'(x) : n \geq 1\}$ is collectively compact whether or not this was assumed. Hence, for each $u \in \mathscr{X}$, there is a subsequence $\{K_{n_i}\}$ and an element $y(x, u) \in \mathscr{X}$ such that

$$K_{n_i}'(x)u \to y(x, u).$$

By the triangle inequality,

$$\|K(x + \delta u) - Kx - \delta y(x, u)\| \leq \|K(x + \delta u) - K_{n_i}(x + \delta u)\|$$
$$+ \|K_{n_i}(x + \delta u) - K_{n_i}x - \delta K_{n_i}'(x)u\|$$
$$+ \|K_{n_i}x - Kx\| + \delta \|K_{n_i}'(x)u - y(x, u)\|.$$

It follows that

$$\left\| \frac{K(x + \delta u) - Kx}{\delta} - y(x, u) \right\| \to 0 \quad \text{as} \quad \delta \to 0.$$

This implies that $y(x, u)$ is independent of the particular subsequence $\{K_{n_i}\}$ which entered into its definition. An application of the result used in the proof of Lemma 5.2 yields

$$K_n'(x)u \to y(x, u)$$

for each $u \in \mathscr{X}$. By a standard argument involving the Banach-Steinhaus theorem, there exists $L_x \in [\mathscr{X}]$ such that

$$K_n'(x) \to L_x, \qquad L_x u = y(x, u),$$
$$\left\| \frac{K(x + \delta u) - Kx}{\delta} - L_x u \right\| \to 0 \quad \text{as} \quad \delta \to 0.$$

So K is differentiable at x, $K'(x) = L_x$ and $K_n'(x) \to K'(x)$.

By Propositions 6.9 and 6.11, the hypothesis that $K_n'(x) \to K'(x)$ in §6.6 follows from the other assumptions.

In the next theorem it is shown that certain pointwise hypotheses hold uniformly over any compact set. An immediate generalization is obtained if

111

K' is replaced by any other function from $\mathscr{X}$ to $[\mathscr{X}]$ with the prescribed properties.

THEOREM 6.12. Let $\mathscr{K}$ be a set of operators on $\mathscr{X}$ and let $\mathscr{V} \subset \mathscr{X}$ be compact. Assume

> each $K \in \mathscr{K}$ differentiable on $\mathscr{V}$,
> $\{K': K \in \mathscr{K}\}$ equicontinuous at each $x \in \mathscr{V}$,
> $\{K'(x): K \in \mathscr{K}\}$ collectively compact for each $x \in \mathscr{V}$.

Then $\{K'(x): K \in \mathscr{K}, x \in \mathscr{V}\}$ is collectively compact.

Proof. Recall that a continuous function from one metric space to another is uniformly continuous on any compact set. For the same reasons, equicontinuity is uniform on any compact set. Thus, $\{K': K \in \mathscr{K}\}$ is uniformly equicontinuous on $\mathscr{V}$: for each $\varepsilon > 0$ there is a $\delta > 0$ such that

$$\|K'(x)u - K'(y)u\| \le \|K'(x) - K'(y)\| < \varepsilon$$

if $K \in \mathscr{K}$, $u \in \mathscr{B}$, $x,y \in \mathscr{V}$ and $\|x - y\| < \delta$.

Since $\mathscr{V}$ is totally bounded, it has a finite δ-net $\mathscr{V}_\delta$. Now

$$\{K'(y): K \in \mathscr{K}, y \in \mathscr{V}_\delta\}\mathscr{B}$$

is a totally bounded ε-net for

$$\{K'(x): K \in \mathscr{K}, x \in \mathscr{V}\}\mathscr{B}.$$

So the latter set is totally bounded and $\{K'(x): K \in \mathscr{K}, x \in \mathscr{V}\}$ is collectively compact.

COROLLARY 6.13. Let K be an operator on $\mathscr{X}$ and let $\mathscr{V} \subset \mathscr{X}$ be compact. Assume K differentiable on $\mathscr{V}$, K' continuous on $\mathscr{V}$, and $K'(x)$ compact for each $x \in \mathscr{V}$. Then $\{K'(x): x \in \mathscr{V}\}$ is collectively compact.

Theorem 6.12 and Corollary 6.13 imply that in the situation of §6.6 the sets

$$\{K'(x): x \in \mathscr{V}\}, \qquad \{K'_n(x): x \in \mathscr{V}, n \ge 1\}$$

are collectively compact.

112

appendix I

BASIC PROPERTIES OF
COMPACT LINEAR OPERATORS

Most of the basic properties of compact linear operators given without proof in the text are derived here. Although an acquaintance with the derivations is not necessary for an understanding of the collectively compact operator approximation theory, it would certainly be beneficial.

Let $\mathscr{X}$ be a real or complex normed linear space. As before, $[\mathscr{X}]$ denotes the space of bounded linear operators $T: \mathscr{X} \to \mathscr{X}$ equipped with the operator norm. Let $\mathscr{N}(T) = \{x \in \mathscr{X}: Tx = 0\}$, the null space of T. A linear operator K on $\mathscr{X}$ is *compact* iff for each bounded set $\mathscr{S} \subset \mathscr{X}$ the set $K\mathscr{S}$ is relatively compact (i.e., $\overline{K\mathscr{S}}$ is compact). Equivalent criteria are that $K\mathscr{S}$ is sequentially compact or, when $\mathscr{X}$ is complete, that $K\mathscr{S}$ is totally bounded. Every compact linear operator K on $\mathscr{X}$ is bounded, hence $K \in [\mathscr{X}]$. Our principal goal is the *Fredholm alternative*: for any compact $K \in [\mathscr{X}]$, $(I - K)^{-1}$ exists (as an operator defined on $(I - K)\mathscr{X}$) iff $(I - K)\mathscr{X} = \mathscr{X}$, in which case $(I - K)^{-1}$ is bounded and, hence, $(I - K)^{-1} \in [\mathscr{X}]$.

We begin with several properties of compact operators which follow quite readily from the definition. The compact linear operators on $\mathscr{X}$ form a subspace of $[\mathscr{X}]$. This subspace is closed when $\mathscr{X}$ is complete. Since the bounded sets in a finite dimensional normed linear space are relatively compact, every $K \in [\mathscr{X}]$ with dim $K\mathscr{X} < \infty$ is compact. Hence, every $K \in [\mathscr{X}]$ is compact when dim $\mathscr{X} < \infty$. If K is compact and T is bounded, then KT and TK are compact. So any finite product of compact operators in $[\mathscr{X}]$ is compact. It follows that, for each compact $K \in [\mathscr{X}]$, the operators,

$$p(K) = a_1 K + a_2 K^2 + \cdots + a_n K^n \qquad (a_i \text{ scalar})$$

are compact. For example, when $K \in [\mathscr{X}]$ is compact, the operators K_n defined by

$$(I - K)^n = I - K_n, \qquad n = 1, 2, \ldots,$$

are compact. This implies that any property of $I - K$ for an arbitrary compact operator $K \in [\mathscr{X}]$ also holds for $(I - K)^n$, $n \geq 1$.

An important tool in the study of compact operators is the Riesz lemma, which is stated without proof.

LEMMA I. Let $\mathscr{M}$ be a closed proper subspace of $\mathscr{X}$. For each $\varepsilon > 0$ there exists $x_\varepsilon \in \mathscr{M}$ such that

$$\|x_\varepsilon\| = 1, \qquad \|x_\varepsilon - y\| \geq 1 - \varepsilon \quad \text{for} \quad y \in \mathscr{M}.$$

If dim $\mathscr{M} < \infty$ there exists $x \in \mathscr{X}$ such that

$$\|x\| = 1, \qquad \|x - y\| \geq 1 \quad \text{for} \quad y \in \mathscr{M}.$$

When $\mathscr{X}$ is infinite dimensional, repeated application of Lemma 1 yields an infinite sequence $\{x_n\}$ such that $\|x_n\| = 1$ for all n and $\|x_m - x_n\| \geq 1$ for $m \neq n$. It follows that every bounded set in $\mathscr{X}$ is relatively compact iff dim $\mathscr{X} < \infty$. Hence the identity operator I on $\mathscr{X}$ is compact iff dim $\mathscr{X} < \infty$. Suppose $K \in [\mathscr{X}]$ is compact and K^{-1} exists, so that $K^{-1}K = I$. Then K^{-1} is bounded iff dim $\mathscr{X} < \infty$.

PROPOSITION 2. Let $K \in [\mathscr{X}]$ be compact. Then dim $\mathscr{N}(I - K) < \infty$. Moreover, dim $\mathscr{N}[(I - K)^n] < \infty$ for each $n \geq 1$.

Proof. Note that $Kx = x$ for each $x \in \mathscr{N}(I - K)$.

PROPOSITION 3. Let $K \in [\mathscr{X}]$ be compact and $\mathscr{S} \subset \mathscr{X}$. Then

$$\mathscr{S} \text{ closed, bounded} \Rightarrow (I - K)\mathscr{S} \text{ closed}.$$

Proof. Let $y \in \overline{(I - K)\mathscr{S}}$. Then there is an infinite sequence $\{x_n\}$ in $\mathscr{S}$ such that

$$(I - K)x_n \to y.$$

Since $K\mathscr{S}$ is sequentially compact, there exist a subsequence $\{x_{n_i}\}$ and $v \in \mathscr{X}$ such that $Kx_{n_i} \to v$. Let $x = y + v$. Then

$$x_{n_i} = (I - K)x_{n_i} + Kx_{n_i} \to x, \qquad x \in \mathscr{S},$$
$$(I - K)x_{n_i} \to (I - K)x,$$
$$(I - K)x = y, \qquad y \in (I - K)\mathscr{S}.$$

Thus, $(I - K)\mathscr{S}$ is closed.

Some additional notation will be needed. Let

$$\mathscr{U} = \{x \in \mathscr{X} : \|x\| = 1\},$$

the unit sphere in $\mathscr{X}$. For $T \in [\mathscr{X}]$ let

$$m(T) = \inf_{x \in \mathscr{U}} \|Tx\| = \inf_{x \neq 0} \frac{\|Tx\|}{\|x\|}.$$

LEMMA 4. Let $T \in [\mathscr{X}]$. Then

a. $\exists T^{-1} \Leftrightarrow 0 \notin T\mathscr{U}$,

b. $\exists T^{-1}$ bounded $\Leftrightarrow 0 \notin \overline{T\mathscr{U}} \Leftrightarrow m(T) > 0 \Rightarrow$

$$\|T^{-1}\| = \frac{1}{m(T)}.$$

c. $\exists T^{-1}, T\mathscr{U}$ closed $\Rightarrow T^{-1}$ bounded, $T\mathscr{X}$ closed.

Proof. Both a and b are easy to establish. Now assume that T^{-1} exists and $T\mathscr{U}$ is closed. From a and b, T^{-1} is bounded. Let $y \in \overline{T\mathscr{X}}$, $y \neq 0$. Then there exists $\{y_n\} \subset T\mathscr{X}$ such that $y_n \to y$ and $y_n \neq 0$ for all n. Let $x_n = T^{-1}y_n$. Then $x_n \neq 0$ for all n, and $\{x_n\}$ is Cauchy. So $\{\|x_n\|\}$ is Cauchy, hence convergent. Suppose $\|x_n\| \to r$. Since T is bounded and $Tx_n = y_n \to y \neq 0$, $r \neq 0$. Now

$$T\left(\frac{rx_n}{\|x_n\|}\right) \to y, \qquad y \in \overline{T(r\mathscr{U})}.$$

But since $T\mathscr{U}$ is closed, $T(r\mathscr{U})$ is closed and $y \in T(r\mathscr{U}) \subset T\mathscr{X}$. Therefore, $T\mathscr{X}$ is closed.

It should be noted that Lemma 4 applies without change to bounded
linear operators from one normed linear space to another. In a and b, the
operator T need not be bounded.

The next result is a direct consequence of Proposition 3 and Lemma 4.

PROPOSITION 5. Let $K \in [\mathscr{X}]$ be compact and $(I - K)^{-1}$ exist.
Then $(I - K)^{-1}$ is bounded.

We could have concluded also that $(I - K)\mathscr{X}$ is closed. However,
this is true even without the assumption that $(I - K)^{-1}$ exists. The proof will
depend on the following auxiliary material.

Let $\mathscr{N}$ be a closed subspace of $\mathscr{X}$. The quotient space

$$\mathscr{X}/\mathscr{N} = \{x + \mathscr{N} : x \in \mathscr{X}\}$$

is a normed linear space with zero element $\mathscr{N}$ and norm

$$\|\hat{x}\| = \inf_{u \in \hat{x}} \|u\|, \qquad \hat{x} \in \mathscr{X}/\mathscr{N}.$$

The quotient map $\Phi: \mathscr{X} \to \mathscr{X}/\mathscr{N}$ is defined by $\Phi x = x + \mathscr{N}$. The operator
Φ is linear, $\Phi\mathscr{X} = \mathscr{X}/\mathscr{N}$, Φ is bounded (hence, continuous), and $\|\Phi\| = 1$.
For each $\hat{x} \in \mathscr{X}/\mathscr{N}$ and each $c > 1$ there exists $x \in \mathscr{X}$ such that

$$\Phi x = \hat{x}, \qquad \|\hat{x}\| \le \|x\| \le c\|\hat{x}\|.$$

If $\mathscr{S} \subset \mathscr{X}/\mathscr{N}$ is bounded and $\|\hat{x}\| \le r < \infty$ for $\hat{x} \in \mathscr{S}$, then for any $c > 1$
the set

$$\mathscr{S} = \{x \in \mathscr{X} : \Phi x \in \mathscr{S}, \quad \|x\| \le cr\}$$

is bounded and $\Phi\mathscr{S} = \mathscr{S}$. If $\mathscr{S}$ is closed, then $\mathscr{S}$ is closed because Φ is
continuous.

Now let $T \in [\mathscr{X}]$ and $\mathscr{N} = \mathscr{N}(T)$. A linear operator $\hat{T}: \mathscr{X}/\mathscr{N} \to \mathscr{X}$ is
defined by

$$\hat{T}(x + \mathscr{N}) = T(x + \mathscr{N}) = Tx.$$

Note that $\hat{T}(\mathscr{X}/\mathscr{N}) = T\mathscr{X}$. Since

$$\hat{T}(x + \mathscr{N}) = 0 \Rightarrow x \in \mathscr{N} \Rightarrow x + \mathscr{N} = \mathscr{N},$$

$\hat{T}^{-1}$ exists. An equivalent definition of $\hat{T}$ is $\hat{T}\Phi = T$. The properties of Φ imply that $\hat{T}$ is bounded and $\|\hat{T}\| = \|T\|$.

LEMMA 6. Let $T \in [\mathscr{X}]$. Assume that, for $\mathscr{S} \subset \mathscr{X}$,

$$\mathscr{S} \text{ closed, bounded} \Rightarrow T\mathscr{S} \text{ closed.}$$

Then $T\mathscr{X}$ is closed and $\hat{T}^{-1}$ is bounded.

Proof. Let $\mathscr{S} \subset \mathscr{X}/\mathscr{N}$ be closed and bounded. As shown above, there is a closed and bounded set $\mathscr{S} \subset \mathscr{X}$ such that $\Phi\mathscr{S} = \mathscr{S}$. Then $\hat{T}\mathscr{S} = \hat{T}\Phi\mathscr{S} = T\mathscr{S}$, so that $\hat{T}\mathscr{S}$ is closed. By Lemma 4c, $\hat{T}^{-1}$ is bounded and $\hat{T}(\mathscr{X}/\mathscr{N}) = T\mathscr{X}$ is closed.

The next result follows from Lemma 6 and Proposition 3.

THEOREM 7. Let $K \in [\mathscr{X}]$ be compact and $T = I - K$. Then $T\mathscr{X}$ is closed and $\hat{T}^{-1}$ is bounded. Moreover, $T^n\mathscr{X}$ is closed for each $n \geq 1$.

In this paragraph and the theorem which follows, $\mathscr{X}$ may be an arbitrary linear space. Norms play no role. Let Θ denote the zero subspace of $\mathscr{X}$ and $\mathscr{L}(\mathscr{X})$ the space of linear operators $T: \mathscr{X} \to \mathscr{X}$. As usual, when the inverse operator exists it is denoted by T^{-1}. For each $T \in \mathscr{L}(\mathscr{X})$ let $T^{\sim 1}$ be the set-function inverse of T:

$$T^{\sim 1}\mathscr{S} = \{x \in \mathscr{X} : Tx \in \mathscr{S}\}, \qquad \mathscr{S} \subset \mathscr{X}.$$

In particular, $T^{\sim 1}\Theta = \mathscr{N}(T)$, the null space of T. Define $T^{\sim n} = (T^n)^{\sim 1}$ for $n = 0, 1, 2, \ldots$. Then $T^{\sim n}\Theta = \mathscr{N}(T^n)$. It follows easily from

$$T^{\sim m}T^{\sim n} = T^{\sim(m+n)} \quad \text{and} \quad T^m T^n = T^{(m+n)}$$

that there exist unique nonnegative integers or $+\infty$, called the *ascent* $\alpha = \alpha(T)$ and the *descent* $\delta = \delta(T)$ of T, such that for $n = 0, 1, 2, \ldots$

$$T^{\sim n}\Theta \subset T^{\sim(n+1)}\Theta \quad \text{with proper inclusion iff} \quad n < \alpha,$$

$$T^n\mathscr{X} \supset T^{n+1}\mathscr{X} \quad \text{with proper inclusion iff} \quad n < \delta.$$

Since $T^{\sim 0} = \Theta$, $\alpha = 0$ iff T^{-1} exists. Since $T^0\mathscr{X} = \mathscr{X}$, $\delta = 0$ iff $T\mathscr{X} = \mathscr{X}$.

THEOREM 8. Let $T \in \mathcal{L}(\mathcal{X})$. Assume $\alpha = \alpha(T) < \infty$ and $\delta = \delta$ $(T) < \infty$. Then $\alpha = \delta$,

$$\mathcal{X} = T^\alpha \mathcal{X} \oplus T^{\sim\alpha}\Theta,$$

$$T(T^\alpha \mathcal{X}) = T^\alpha \mathcal{X}, \qquad T(T^{\sim\alpha}\Theta) \subset T^{\sim\alpha}\Theta,$$

T is an isomorphism on $T^\alpha \mathcal{X}$, T is nilpotent of degree α on $T^{\sim\alpha}\Theta$, and

$$\exists T^{-1} \Leftrightarrow T\mathcal{X} = \mathcal{X}.$$

Proof. For $x \in \mathcal{X}$, $\mathcal{S} \subset \mathcal{X}$, and $m \geq 0$,

$$T^{\sim 1}Tx = x + T^{\sim 1}\Theta,$$

$$T^{\sim 1}T\mathcal{S} = \mathcal{S} + T^{\sim 1}\Theta,$$

$$T\mathcal{S} = T\mathcal{X} \Leftrightarrow \mathcal{X} = \mathcal{S} + T^{\sim 1}\Theta,$$

$$T^m \mathcal{S} = T^m \mathcal{X} \Leftrightarrow \mathcal{X} = \mathcal{S} + T^{\sim m}\Theta.$$

For any $m \geq 0$ and any $n \geq 1$, the definition of δ yields

$$\delta \leq m \Leftrightarrow T^m(T^n \mathcal{X}) = T^m \mathcal{X} \Leftrightarrow \mathcal{X} = T^n \mathcal{X} + T^{\sim m}\Theta.$$

It follows that

$$\mathcal{X} = T^n \mathcal{X} + T^{\sim\delta}\Theta, \qquad n \geq 0.$$

From the definition of α, $T^{\sim\alpha}\Theta \supset T^{\sim\delta}\Theta$ and $\mathcal{X} = T^n \mathcal{X} + T^{\sim\alpha}\Theta$ for $n \geq 0$. Hence, $\delta \leq \alpha$.

For $\mathcal{S} \subset \mathcal{X}$ and $m \geq 0$,

$$TT^{\sim 1}\mathcal{S} = T\mathcal{X} \cap \mathcal{S},$$

$$T^{\sim 1}\mathcal{S} = T^{\sim 1}\Theta \Leftrightarrow T\mathcal{X} \cap \mathcal{S} = \Theta,$$

$$T^{\sim m}\mathcal{S} = T^{\sim m}\Theta \Leftrightarrow T^m \mathcal{X} \cap \mathcal{S} = \Theta.$$

For any $m \geq 0$ and any $n \geq 1$, the definition of α yields

$$\alpha \leq m \Leftrightarrow T^{\sim m}T^{\sim n}\Theta = T^{\sim m}\Theta \Leftrightarrow T^m \mathcal{X} \cap T^{\sim n}\Theta = \Theta.$$

It follows that

$$T^\alpha \mathcal{X} \cap T^{\sim n}\Theta = \Theta, \qquad n \geq 0.$$

From the definition of δ, $T^\delta \mathcal{X} \subset T^\alpha \mathcal{X}$, $T^\delta \mathcal{X} \cap T^{\sim n}\Theta = \Theta$ for $n \geq 0$, and $\alpha \leq \delta$. Thus, $\alpha = \delta$.

Let $\delta = \alpha$ and $n = \alpha$ above to obtain $\mathcal{X} = T^\alpha \mathcal{X} + T^{\sim\alpha}\Theta$ and $T^\alpha \mathcal{X} \cap T^{\sim\alpha}\Theta = \Theta$. Thus, $\mathcal{X} = T^\alpha \mathcal{X} \oplus T^{\sim\alpha}\Theta$. Let $n = 1$ above to obtain

$T^\alpha \mathcal{X} \cap T^{\sim 1}\Theta = \Theta$, which implies that the restriction of T to $T^\alpha \mathcal{X}$ is one-to-one. From $\alpha = \delta$ and the definition of δ, $T(T^\alpha \mathcal{X}) = T^{\alpha+1}\mathcal{X} = T^\alpha \mathcal{X}$. Since $T^\alpha x = 0 \Rightarrow T^\alpha(Tx) = 0$, $T(T^{\sim \alpha}\Theta) \subset T^{\sim \alpha}\Theta$. Obviously $T^\alpha x = 0$ for $x \in T^{\sim \alpha}\Theta$. But $T^{\sim(\alpha-1)}\Theta \ne T^{\sim \alpha}\Theta$. So T is nilpotent of degree α on $T^{\sim \alpha}\Theta$. Finally,

$$\exists T^{-1} \Leftrightarrow \alpha = 0 \Leftrightarrow \delta = 0 \Leftrightarrow T\mathcal{X} = \mathcal{X}.$$

Again let $\mathcal{X}$ be a normed linear space.

THEOREM 9. For any compact $K \in [\mathcal{X}]$, $\alpha(I - K) = \delta(I - K) < \infty$.

Proof. Since $\mathcal{N}[(I - K)^{n-1}] \subsetneq \mathcal{N}[(I - K)^n]$ for each integer n with $1 \le n \le \alpha(I - K)$, Lemma 1 and Proposition 2 yield $x_n \in \mathcal{N}[(I - K)^n]$ such that

$$\|x_n\| = 1, \qquad \|x_n - v\| \ge 1 \quad \text{for} \quad v \in \mathcal{N}[(I - K)^{n-1}].$$

For any integers m and n with $1 \le m < n \le \alpha(I - K)$,

$$Kx_n - Kx_m = x_n - v,$$
$$v = x_m - (I - K)x_m + (I - K)v_n \in \mathcal{N}[(I - K)^{n-1}],$$
$$\|Kx_n - Kx_m\| \ge 1.$$

Since K is compact, it follows that $\alpha(I - K) < \infty$. A similar argument proves $\delta(I - K) < \infty$. So $\alpha(I - K) = \delta(I - K)$ by Theorem 8.

The next theorem comprises the Fredholm alternative in an augmented form. Recall that the codimension of a subspace $\mathcal{M} \subset \mathcal{X}$ is the dimension $n \le \infty$ of a subspace $\mathcal{N} \subset \mathcal{X}$ such that $\mathcal{X} = \mathcal{M} \oplus \mathcal{N}$, and that n is independent of the choice of $\mathcal{N}$.

THEOREM 10. For any compact $K \in [\mathcal{X}]$,

a. $\exists(I - K)^{-1} \Leftrightarrow (I - K)\mathcal{X} = \mathcal{X} \Rightarrow (I - K)^{-1} \in [\mathcal{X}]$,

b. $\dim \mathcal{N}(I - K) = \text{codim}\,(I - K)\mathcal{X}$.

Proof. Proposition 5 and Theorems 8 and 9 yield a. Let $n(K)$ be the minimum of the two numbers in b. By Proposition 2, $n(K)$ is finite. When $n(K) = 0$, a implies b. Let $n(K) \geqq 1$ and choose nonzero elements $u \in \mathcal{N}(I - K)$ and $v \notin (I - K)\mathcal{X}$. By the Hahn–Banach theorem, there is a bounded linear functional φ on $\mathcal{X}$ such that $\varphi(u) \neq 0$. Define $L \in [\mathcal{X}]$ by $Lx = \varphi(x)v$. Then L is compact and $n(K + L) = n(K) - 1$. Induction yields b.

Let $K \in [\mathcal{X}]$ and $\lambda \neq 0$. Since $(\lambda I - K) = \lambda(I - \lambda^{-1}K)$, all of the results obtained for $I - K$ apply to $\lambda I - K$. By Theorem 10 and Proposition 2, either $(\lambda I - K)^{-1} \in [\mathcal{X}]$ or λ is an eigenvalue of K, in which case the eigenmanifold $\mathcal{N}(\lambda I - K)$ and the generalized eigenmanifolds $\mathcal{N}[(\lambda I - K)^n]$ are finite dimensional. The number $\lambda = 0$ is exceptional. Either 0 is an eigenvalue of K or K^{-1} exists, but K^{-1} is bounded only when dim $\mathcal{X} < \infty$.

THEOREM II. For any compact $K \in [\mathcal{X}]$, the eigenvalues of K form either a finite set or an infinite sequence which converges to zero.

Proof. Fix $\varepsilon > 0$ arbitrarily. Let K have at least N distinct eigenvalues λ_n. $1 \leq n \leq N$, with $|\lambda_n| \geqq \varepsilon$. Choose corresponding eigenvectors x_n. By an elementary argument $\{x_n : 1 \leqq n \leqq N\}$ is linearly independent. For $1 \leqq n \leqq N$ let $\mathcal{X}_n$ be the subspace spanned by $\{x_m : 1 \leqq m \leqq n\}$. By Lemma 1 there exist $u_n \in \mathcal{X}_n$, $1 \leqq n \leqq N$ such that

$$\|u_n\| = 1, \qquad \|u_n - v\| \geqq 1 \quad \text{for} \quad v \in \mathcal{X}_{n-1}.$$

Then $Ku_n \in \mathcal{X}_n$, $(\lambda_n I - K)u_n \in \mathcal{X}_{n-1}$ and, for $1 \leqq m < n \leqq N$,

$$Ku_n - Ku_m = \lambda_n(u_n - v), \qquad v = \lambda_n^{-1}[(\lambda_n I - K)u_n + Ku_m] \in \mathcal{X}_{n-1},$$

$$\|Ku_n - Ku_m\| = |\lambda_n| \, \|u_n - v\| \geqq \varepsilon.$$

Since K is compact, it follows that K has only a finite number of eigenvalues λ with $|\lambda| \geqq \varepsilon > 0$, which implies the desired result.

This concludes our summary of basic properties of compact operators. For further information, see Dunford and Schwartz [38] or Taylor [70].

NUMERICAL TREATMENT OF INTEGRAL EQUATIONS

Joel Davis

This appendix provides organization for the practical implementation of some of the methods of the text for solving linear Fredholm equations of the second kind. Numerical results from several sources will illustrate the utility of the various methods.

We wish to describe the organization of a system to invert Fredholm operators in a manner useful to the user at any installation; no actual programming will be presented. The system consists of a main subroutine called FRED and three *modules* or sets of subroutines that are used by FRED to delineate the particular problem, quadrature rule, and matrix processor, respectively. Several versions of FRED and its modules are available, emphasizing particular compromises of speed and rigor.

The subroutine FRED solves a particular problem producing sample values of the answer function at given test points (not necessarily the quadrature points). Input parameters include the number of quadrature points, the number and position of the test points, and possible parameters used in defining the kernel and right-hand side of the equation. Output parameters include the answer function evaluated at the test points, estimates of error and inverse norms, condition number for the matrix involved, and an error code to delineate possible failures. The method used is indicated in §2.4. of Chapter 2.

The problem module consists of function subroutines that define k,y, and (optionally) the discontinuities of k and its partial derivatives. In a rigorous calculation additional function subroutines are needed for estimating the error in the above routines. This module is supplied by the user, although a few demonstration problems (with known answers included) are available.

The quadrature rule consists of two subroutines. The first is called by FRED to define the quadrature points, and in some cases, the quadrature weights. The second is a function subroutine which implements $\hat{K}_n$. This latter uses $k(s,t)$ from the problem module. Several quadrature rules are available.

The matrix processor solves linear systems and is used by FRED to solve

$$(I_n - \tilde{K}_n)v_n = P_n y.$$

For fast, nonrigorous solutions a simple Gauss routine is used. For slower, rigorous solutions a combination of Gauss elimination and the Hotelling procedure is used. In the latter case, besides the answer vector, output includes necessary norms, e.g., $\|(I_n - \tilde{K}_n)^{-1}\|$.

In a system run, a user includes FRED, a problem module, a particular quadrature rule, and matrix processor. Except for the problem module, which the user usually provides, the modules and FRED would be available as convenient, precompiled procedures.

The above system was partially implemented, in the PL1 language, on an IBM-360-75 at the University of Illinois, with computer time courtesy of their Computer Science Department. Examples of results from three runs are presented in Tables 1–3. The uniform norm is used in all cases. Condition numbers for the matrices used are indicated.

Problem 1 is typical of problems involving smooth kernels and solutions. Note that 10-point Gauss is at least a factor of 10^{13} times better than the 11 point trapezoidal rule.

Problem 2 is a typical Green's function kernel resulting from a boundary value problem. The first partial derivatives are discontinuous. Removal of the singularity improves the results in a striking way.

Problem 3 results from a Volterra integral equation. Note that k, itself, is discontinuous. Singularity removal is extremely important here.

122

In these three problems the condition number is modest and deteriorates only slightly as n increases.

Problems 4 and 5 are from Yungen [76]. His analysis (partly automated) took into account all sources of error. The key quantities $\|K_n K - K^2\|$, $\|K\|$, $\|y\|$, and $\|K_n y - Ky\|$ were, at the option of the user, handled in one of two manners. In the *rigorous* option, these quantities were calculated analytically using well known quadrature error formulas. In the second approach these quantities were *estimated* approximately by the program using a high order quadrature rule. This option is slower (in machine time), not rigorous, but gives more realistic estimates. Finally, in these examples the exact solution was known so that the exact error could be found. Rounding error made a very small contribution to the estimates.

An alternative method was tested by Yungen. Suppose that $w = Ky$ can be calculated analytically. Then, using the above technique we solve $(I - K)z = w$. The solution to the original problem is $z + w$. The results of this approach are significantly better than the direct approach in the examples tested (cf. §3.7 in Chapter 3).

The last example (Table 6) is due to Atkinson [19]. The unbounded singularity along $s = t$ is removed by factorization (see §3.3 in Chapter 3). Some such transformation is essential in this problem to get any results whatsoever. For examples on the interval $[0,\infty)$ see a further paper by Atkinson [22].

TABLE I

Problem 1: $k(s,t) = 10e^{st}$ $\qquad\qquad\qquad\qquad\qquad\qquad 0 \leqq s,t \leqq 1$

$\qquad\qquad y(s) = s - 10[(s - 1)e^s + 1]/s^2$

$\qquad\qquad x(s) = s \qquad \|x\| = 1$

Method: n-point trapezoidal rule, $\qquad\qquad\qquad\qquad\qquad n = 11, 21, 41, 81$

$\qquad\qquad \|x - x_n\| \leqq \dfrac{24.2}{(n - 1)^2}$

$\qquad\qquad \|x - x_{11}\| \approx .18$

$\qquad\qquad$ Condition ≈ 398 $\qquad\qquad\qquad\qquad\qquad$ if $n = 11$

Method: n-point Simpson's rule, $\qquad\qquad\qquad\qquad\qquad n = 11, 21, 41, 81$

$\qquad\qquad \|x - x_n\| \leqq \dfrac{2.81}{(n - 1)^4}$

$\qquad\qquad \|x - x_{11}\| \approx .00028$

$\qquad\qquad$ Condition ≈ 568 $\qquad\qquad\qquad\qquad\qquad$ if $n = 11$

Method: 10-point Gauss

$\qquad\qquad \|x - x_{10}\| \leqq \tfrac{1}{2} \times 10^{-13}$

$\qquad\qquad$ Condition ≈ 534

Machine: IBM-360-75, double precision

124

TABLE 2

Problem 2: $k(s,t) = 10t(1 - s)$ $0 \leqq t \leqq s \leqq 1$

$k(s,t) = 10s(1 - t)$ $0 \leqq s \leqq t \leqq 1$

$y(s) = \left(1 - \dfrac{10}{\pi^2}\right) \sin \pi s$

$x(s) = \sin \pi s$

$\|x\| = 1$

Method: n-point trapezoidal rule, $n = 11, 21, 41, 81$

$\|x - x_n\| \leqq \dfrac{64}{(n - 1)^2}$

$\|x - x_{21}\| \approx .14,$ Condition ≈ 209

Method: n-point Simpson's rule, $n = 11, 21, 41, 81$

$\|x - x_n\| \leqq \dfrac{85}{(n - 1)^2}$

$\|x - x_{21}\| \approx .18,$ Condition ≈ 208

Method: n-point repeated 10-point Gauss, $n = 10, 20, 40, 80$

$\|x - x_n\| \leqq \dfrac{94}{n^2}$

$\|x - x_{21}\| \approx .20,$ Condition ≈ 212

Method: n-point Simpson's rule modified to remove singularity $n = 11, 21, 41, 81$
(See §3.4 in Chapter 3)

$\|x - x_n\| \leqq \dfrac{125}{(n - 1)^4}$

$\|x - x_{21}\| \approx .00078,$ Condition ≈ 252

Machine: IBM-360-75, double precision

TABLE 3

Problem 3: $k(s,t) = \sin s \cos t$ $0 \leqq t \leqq s \leqq 1$

            $k(s,t) = 0$ $0 \leqq s < t \leqq 1$

            $y(s) = \sin s + (1 - \sin s)e^{\sin s}$

            $x(s) = e^{\sin s}$ $\|x\| = e^{\sin 1} \approx 2.32$

Method: n-point Simpson's rule, $n = 11, 21, 41, 81$

$$\|x - x_n\| \leqq \frac{.953}{n - 1}$$

        $\|x - x_{21}\| \approx .038,$ Condition ≈ 4

Method: n-point repeated 10-point Gauss, $n = 10, 20, 40, 80$

$$\|x - x_n\| \leqq \frac{.923}{n}$$

        $\|x - x_{20}\| \approx .045,$ Condition ≈ 4

Method: n-point trapezoidal rule with singularities removed $n = 11, 21, 41, 81$
        (See §3.4 in Chapter 3)

$$\|x - x_n\| \leqq \frac{.187}{(n - 1)^2}$$

        $\|x - x_{21}\| \approx .00047,$ Condition ≈ 4

Method: n-point Simpson's rule, with singularities removed $n = 11, 21, 41, 81$
        (See §3.4 in Chapter 3)

$$\|x - x_n\| < \frac{.085}{(n - 1)^4}$$

        $\|x - x_{21}\| \approx 5.3 \times 10^{-7},$ Condition ≈ 4

Machine: IBM-360-75, double precision

TABLE 4

Problem 4: $k(s,t) = \tfrac{1}{2}e^{s-t}$ $\qquad\qquad\qquad\qquad\qquad\qquad$ $0 \leqq s,t \leqq 1$

$\qquad\qquad$ $y(s) = 1$

$\qquad\qquad$ $x(s) = 1 + e^s - e^{s-1}$ $\qquad$ $\|x\| = e$

Method: $\quad$ 15-point repeated midpoint rule

$\qquad\qquad$ $\|x - x_{15}\| \leqq 7.6 \times 10^{-2}$ $\qquad\qquad\qquad$ (rigorous)

$\qquad\qquad\qquad\qquad\;\; \leqq 5.3 \times 10^{-4}$ $\qquad\qquad\qquad$ (estimate)

$\qquad\qquad\qquad\qquad\;\; \leqq 3.0 \times 10^{-4}$ $\qquad\qquad\qquad$ (actual)

Method: $\quad$ 15-point repeated 5-point Chebychev rule

$\qquad\qquad$ $\|x - x_{15}\| \leqq 6.6 \times 10^{-4}$ $\qquad\qquad\qquad$ (rigorous)

$\qquad\qquad\qquad\qquad\;\; \leqq 3.1 \times 10^{-8}$ $\qquad\qquad\qquad$ (estimate)

$\qquad\qquad\qquad\qquad\;\; \leqq 3.0 \times 10^{-8}$ $\qquad\qquad\qquad$ (actual)

Machine: $\quad$ CDC-3300, ordinary precision

TABLE 5

Problem 5: $k(s,t) = t(1 - s)$ $\qquad\qquad\qquad\qquad\qquad$ $0 \leqq t \leqq s \leqq 1$

$\qquad\qquad$ $k(s,t) = s(1 - t)$ $\qquad\qquad\qquad\qquad\qquad$ $0 \leqq s \leqq t \leqq 1$

$\qquad\qquad$ $y(s) = \tfrac{1}{2}s(1 - s)$

$\qquad\qquad$ $x(s) = (\tan \tfrac{1}{2}) \sin s + \cos s - 1$

$\qquad\qquad$ $\|x\| = \sec \tfrac{1}{2} - 1 \approx .14$

Method: $\quad$ 15-point repeated midpoint rule

$\qquad\qquad$ $\|x - x_{15}\| \leqq 1.45 \times 10^{-2}$ $\qquad\qquad\qquad$ (rigorous)

$\qquad\qquad\qquad\qquad\;\; \leqq 8.4 \times 10^{-5}$ $\qquad\qquad\qquad$ (estimate)

$\qquad\qquad\qquad\qquad\;\; \leqq 1.5 \times 10^{-4}$ $\qquad\qquad\qquad$ (actual)

Method: $\quad$ 15-point repeated 5-point Chebychev rule

$\qquad\qquad$ $\|x - x_{13}\| \leqq 1.47 \times 10^{-2}$ $\qquad\qquad\qquad$ (rigorous)

$\qquad\qquad\qquad\qquad\;\; \leqq 1.1 \times 10^{-4}$ $\qquad\qquad\qquad$ (estimate)

$\qquad\qquad\qquad\qquad\;\; \leqq 2.0 \times 10^{-4}$ $\qquad\qquad\qquad$ (actual)

Time for both methods: $\quad$ 11 seconds
Machine: $\quad$ CDC-3300 ordinary precision

TABLE 6

Problem 6: $k(s,t) = \log |\cos s - \cos t|$ $0 \leqq s,t \leqq 1$

             $y(s) = 1$

             $\|x\| = x(s) = \dfrac{1}{1 + \pi \log 2} \approx .315$

Method: n-point trapezoidal rule with singularity factored $n = 5, 9, 17, 33$
             (See §3.3 in Chapter 3)

             $\|x - x_n\| \leqq \dfrac{.041}{(n-1)^2}$

             $\|x - x_{17}\| \approx .00016$

Method: n-point Simpson's rule with singularity factored $n = 5, 9, 17, 33$
             (See §3.3 in Chapter 3)

             $\|x - x_n\| \leqq \dfrac{.0048}{(n-1)^4}$

             $\|x - x_{17}\| \approx 7.3 \times 10^{-8}$

Machine: CDC-1604, ordinary precision

BIBLIOGRAPHY

1. Anselone, P. M., Integral Equations of the Schwarzschild-Milne Type, J. Math. Mech., 7 (1958), pp. 557–570.

2. ——, Convergence of the Wick-Chandrasekhar Approximation Technique in Radiative Transfer, Astrophysical J., 128 (1958), pp. 124–129.

3. ——, Convergence of the Wick-Chandrasekhar Approximation Technique in Radiative Transfer II, Astrophysical J., 130 (1959), pp. 881–883.

4. ——, Convergence of Chandrasekhar's Method for the Problem of Diffuse Reflection, Monthly Notices Royal Astron. Soc., 120 (1960), pp. 498–503.

5. ——, Convergence of Chandrasekhar's Method for Inhomogeneous Transfer Problems, J. Math. Mech., 10 (1961), pp. 537–546.

6. ——, Convergence and Error Bounds for Approximate Solutions of Integral and Operator Equations, *in* Error in Digital Computation, vol. 2, ed. by L. B. Rall, John Wiley & Sons, Inc., New York, 1965, pp. 231–252.

129

7. Anselone, P. M., Uniform Approximation Theory for Integral Equations with Discontinuous Kernels, SIAM J. Num. Anal., 4 (1967), pp. 245–253; reprinted in Studies in Numerical Analysis I, Soc. Indust. Appl. Math., Philadelphia, 1968.

8. ——, Collectively Compact Approximations of Integral Operators with Discontinuous Kernels, J. Math. Anal. Appl., 22 (1968), pp. 582–590.

9. ——, Collectively Compact and Totally Bounded Sets of Linear Operators, J. Math. and Mech., 17 (1968), pp. 613–622.

10. ——, Compactness Properties of Sets of Operators and Their Adjoints, Math. Zeit., 113 (1970), pp. 233–236.

11. ——, Abstract Riemann Integrals, Monotone Approximations, and Generalizations of Korovkin's Theorem, Tagung über Numerische Methoden der Approximationstheorie, ISNM vol. 15, Birkhaüser, Basel, 1971.

12. —— and J. M. Gonzalez-Fernandez, Uniformly Convergent Approximate Solutions of Fredholm Integral Equations, J. Math. Anal. Appl., 10 (1965), pp. 519–536.

13. —— and R. H. Moore, Approximate Solutions of Integral and Operator Equations, J. Math. Anal. Appl., 9 (1964), pp. 268–277.

14. —— and R. H. Moore, An Extension of the Newton-Kantorovič Method for Solving Nonlinear Equations, with an Application to Elasticity, J. Math. Anal. Appl., 13 (1966), pp. 476–500.

15. —— and T. W. Palmer, Collectively Compact Sets of Linear Operators, Pacific J. Math., 25 (1968). pp. 417–422.

16. —— and T. W. Palmer, Spectral Analysis of Collectively Compact, Strongly Convergent Operator Sequences, Pacific J. Math., 25 (1968), pp. 423–431.

17. —— and T. W. Palmer, Special Properties of Collectively Compact Sets of Linear Operators, J. Math. and Mech., 17 (1968), pp. 853–860.

130

18. Antosiewicz, H. A. and W. C. Rheinboldt, Numerical Analysis and Functional Analysis, *in* A Survey of Numerical Analysis, ed. by J. Todd, McGraw-Hill Book Company, New York, 1962.

19. Atkinson, K. E., The Numerical Solution of Fredholm Integral Equation of the Second Kind, SIAM J. Num. Anal. 4 (1967), pp. 337–348.

20. Atkinson, K. E., The Solution of Non-Unique Linear Integral Equations, Num. Math., 10 (1967), pp. 117–124.

21. ——, The Numerical Solution of the Eigenvalue Problem for Compact Integral Operators, Trans. Amer. Math. Soc., 129 (1967), pp. 458–465.

22. ——, The Numerical Solution of Integral Equations on the Half-Line, SIAM J. Num. Anal. 6 (1969), pp. 375–397.

23. ——, An Iterative Method for the Numerical Solution of Fredholm Linear Integral Equations of the Second Kind, to appear.

24. Baluev, A. N., Approximate Solutions of Non-Linear Integral Equations, Leningr. Gos. Univ. Uch. Zap. Ser. Mat. Nauk., 33 (1958), pp. 28–31.

25. Brakhage, H., Über die Numerische Behandlung von Integralgleichungen nach der Quadraturformelmethode, Num. Math., 2 (1960), pp. 183–196.

26. ——, Zur Fehlerabschätzung für die Numerische Eigenwertbestimmung bei Integralgleichungen, Num. Math., 3 (1961), pp. 174–179.

27. Bryan, C. A., Approximate Solutions to Nonlinear Integral Equations, SIAM J. Num. Anal., 5 (1968), pp. 151–155.

28. Bückner, H., Konvergenzuntersuchungen bei einem Algebraischen Verfahren zur Näherungsweisen Lösung von Integralgleichungen, Math. Nachr., 3 (1950), pp. 358–372.

29. ——, Die Praktische Behandlung von Integralgleichungen, Ergebnisse der Angewandte Mathematik, Band 1, Springer-Verlag, Berlin, 1952.

30. Bückner, H., Numerical Methods for Integral Equations, *in* A Survey of Numerical Analysis, ed. by J. Todd, McGraw-Hill Book Company, New York, 1962.

31. Chandrasekhar, S., Radiative Transfer, Oxford University Press, 1950.

32. Daniel, J. W., Collectively Compact Sets of Gradient Mappings, Nederl. Akad, Wetensch. Proc. Ser. A, 71 (1968), pp. 270–279.

33. Davis, J., The Solution of Nonlinear Operator Equations with Critical Points, Technical Report No. 25, Oregon State University, Corvallis, 1966.

34. —— and J. J. Jacoby, Numerical Integration of Linear Integral Equations with Mild Discontinuity, Num. Math., 13 (1969), pp. 357–361.

35. Dennis, J. E., Jr., On Newton-Like Methods, Numer. Math., 11 (1968), pp. 324–330.

36. De Pree, J. D. and J. A. Higgins, Collectively Compact Sets of Linear Operators, Math. Z., 115 (1970), pp. 366–370.

37. Dieudonné, J., Foundations of Modern Analysis, Academic Press, Inc., New York, 1960.

38. Dunford, N. and J. T. Schwartz, Linear Operators, Part I: General Theory, Interscience Publishers, Inc., New York, 1958.

39. Fox, L. and E. T. Goodwin, The Numerical Solution of Non-Singular Linear Integral Equations, Phil. Trans. Roy. Soc. London, A245 (1953), pp. 501–534.

40. Fredholm, I., Sur une Nouvelle Méthode pour la Résolution du Problème de Dirichlet, Kong. Vetenskaps-Akademiens Förh. Stockholm (1900), pp. 39–46.

41. ——, Sur une Classe d'Équations Fonctionnelles, Acta. Math., (1903), pp. 365–390.

132

42. Hilbert, D., Grundzüge einer Allgemeinem Theorie der Linearen Integralgleichungen, Leipzig, 1912.

43. Hildebrand, F. B., Introduction to Numerical Analysis, McGraw-Hill Book Company, New York, 1956.

44. Hopf, E., Mathematical Problems of Radiative Equilibrium, Cambridge University Press, 1934.

45. James, R. L., Convergence of Positive Operators, Ph.D. Thesis, Oregon State University, Corvallis, 1970.

46. Kantorovich, L. V., Functional Analysis and Applied Mathematics, Uspehi, Mat. Nauk., 3 (1948), pp. 89–185.

47. —— and G. P. Akilov, Functional Analysis in Normed Spaces, trans. by D. E. Brown, The Macmillan Company, New York, 1964.

48. —— and V. I. Krylov, Approximate Methods of Higher Analysis, trans. by C. D. Benster, Interscience Publishers, Inc., New York, 1958.

49. Kato, T., Perturbation Theory for Linear Operators, Springer-Verlag, New York, 1966.

50. Klee, V., Two Renorming Constructions Related to a Question of Anselone, Studia Math., 33 (1969), pp. 231–242.

51. Krasnosel'skii, M. A., Topological Methods in the Theory of Nonlinear Integral Equations, trans. by A. H. Armstrong, The Macmillan Company, New York, 1964.

52. —— and Y. B. Rutickii, Some Approximate Methods of Solving Non-linear Operator Equations Based on Linearization, Sov. Math., 2 (1961), pp. 1542–1546.

53. Milne, W. E., Numerical Calculus, Princeton University Press, Princeton, N.J., 1950.

54. Mysovskih, I. P., Estimation of Error Arising in the Solution of an Integral Equation by the Method of Mechanical Quadratures, Vestnik Leningrad Univ., 11 (1956), pp. 66–72.

55. ——, An Error Estimate for the Numerical Solution of a Linear Integral Equation, Dokl. Akad. Nauk SSSR, 140 (1961), pp. 763–765.

56. ——, On the Method of Mechanical Quadrature for the Solution of Integral Equations, Vestnik Leningrad Univ., 17 (1962), pp. 78–88.

57. Mysovskih, I. P., On Convergence of Newton's Method, Trudy Mat. Inst. Steklov, 28 (1949), pp. 145–147.

58. ——, On the Convergence of the Method of L. V. Kantorovič for the Solution of Nonlinear Functional Equations and Its Applications, Vestnik Leningrad Univ., 11 (1953), pp. 25–48.

59. Moore, R. H., Newton's Method and Variations, in Nonlinear Integral Equations, ed. by P. M. Anselone, University of Wisconsin Press, Madison, 1964.

60. ——, Differentiability and Convergence for Compact Nonlinear Operators, J. Math. Anal. Appl., 16 (1966), pp. 65–72.

61. ——, Approximations to Nonlinear Operator Equations and Newton's Method, Numer. Math., 12 (1968), pp. 23–34.

62. Nestell, M. K., The Convergence of the Discrete Ordinates Method for Integral Equations of Anisotropic Radiative Transfer, Technical Report No. 23, Oregon State University, Corvallis, 1965.

63. Nyström, E. J., Über die Praktische Auflösung von Integralgleichungen mit Anwendungen auf Randwertaufgaben, Acta Math., 54 (1930), pp. 185–204.

64. Ostrowski, A. M., General Existence Criteria for the Inverse of an Operator, Amer. Math. Monthly, 74 (1967), pp. 826–827.

134

65. Palmer, T. W., Totally Bounded Sets of Precompact Linear Operators, Proc. Amer. Math. Soc., 20 (1969), pp. 101–106.

66. Papadopoulos, M., Diffraction by a Refracting Wedge, Tech. Report 297, Mathematics Research Center, U.S. Army, University of Wisconsin, Madison, Wisconsin, 1962.

67. Putnam, C. R., Perturbations of Bounded Operators, Nieuw Archief voor Wiskunde (3), XV (1967), pp. 146–152.

68. Rall, L. B., Computational Solution of Nonlinear Operator Equations, John Wiley & Sons, Inc., New York, 1969.

69. Szegö, G., Orthogonal Polynomials, rev. ed., Colloquium Publ. no. 23, American Mathematical Society, New York, 1959.

70. Taylor, A. E., Introduction to Functional Analysis, John Wiley & Sons, Inc., New York, 1958.

71. Tricomi, F. G., Integral Equations, Interscience Publishers Inc., New York, 1957.

72. Vala, K., On Compact Sets of Compact Operators, Ann. Acad. Sci. Fenn. Ser. Al., 351 (1964), pp. 1–9.

73. Weyl, H., Über die Gleichverteilung der Zahlen mod. Eins., Math. Annalen, 77 (1916), pp. 313–352.

74. Wielandt, H., Error Bounds for Eigenvalues of Symmetric Integral Equations, Proc. Symposium Appl. Math. VI, American Mathematical Society, New York, 1956.

75. Yosida, K., Functional Analysis, Springer-Verlag, Berlin, 1965.

76. Yungen, W. A., Rigorous Computer Inversion of Some Linear Operators, M.S. Thesis, Oregon State University, Corvallis, 1968.

77. Gilbert, R. P., Integral Operator Methods for Approximating Solutions of Dirichlet Problems, with an appendix by K. E. Atkinson. Conference on "Numerische der Approximationstheorie," at Oberwolfach, Germany, June 8–14, 1969.

78. —— and D. L. Colton, On the Numerical Treatment of Partial Differential Equations by Function Theoretic Methods. Symposium on the Numerical Solution of Partial Differential Equations, University of Maryland, May 11–15, 1970.

79. Atkinson, K. E., The Numerical Solution of Fredholm Integral Equations of the Second Kind with Singular Kernels.

136

INDEX

Index

G

Green's function, 41

H

Hahn-Banach theorem, 88
Hilbert space, 82–88

I

Index of an operator, 70
Integral equation, Fredholm, 13
  Urysohn, 93
Integral operator, Fredholm, 14
  Volterra, 38
  Urysohn, 94
Inverse operator, 1–12

K

Kantorovich theorem, 99
  variations, 102
Kernel, continuous, 18
  discontinuous, 28
  finite rank, 56
  mildly discontinuous, 31
  singular, 35, 52, 55
  Volterra, 30, 38
Korovkin theorem, 28

L

Linear functional, 15, 23

N

Neumann problem, 44
Neumann series, 2
Newton's method, 98
  modified, 100
Nonlinear elasticity, 101
Numerical-integral operator, 18, 94

Numerical solutions of integral equations, 121–128

P

Partial differential equation, 45
Practical homotopy, 100
Projection, 68
  orthogonal, 82

Q

Quadrature formula, 15, 23

R

Regular set, 26
Relatively compact set, 4
Resolvent, 60
  equation, 61
  operator, 9
  set, 58
  set, extended, 60
Riemann integral, 22
Riesz lemma, 67

S

Sequentially compact set, 4
Spectral set, 72
  projection, 73
  subspace, 73
Spectrum, 58
  point, 58

T

Totally bounded set, 4, 86
  of operators, 81
Transport equation, 51

U

Uniform boundedness principle, 7

138